Lecture Notes in Social Networks

More information about this series at http://www.springer.com/series/8768

Tansel Özyer • Sambit Bakshi • Reda Alhajj
Editors

Social Networks and Surveillance for Society

Springer

Editors
Tansel Özyer
Department of Computer Engineering
TOBB University of Economics
and Technology
Ankara, Turkey

Sambit Bakshi
Department of Computer Engineering
National Institute of Technology Rourkela
Rourkela, Odisha, India

Reda Alhajj
Department of Computer Science
University of Calgary
Calgary, AB, Canada

ISSN 2190-5428 ISSN 2190-5436 (electronic)
Lecture Notes in Social Networks
ISBN 978-3-030-08657-2 ISBN 978-3-319-78256-0 (eBook)
https://doi.org/10.1007/978-3-319-78256-0

Printed on acid-free paper

This Springer imprint is published by the registered company Springer International Publishing AG part of Springer Nature.
The registered company address is: Gewerbestrasse 11, 6330 Cham, Switzerland

Preface

The study of social networks emerged in the nineteenth century and concentrated on small networks of humans and organizations. However, the development of Web 2.0 tremendously influenced the field and shifted the concentration from social sciences into multidisciplinary approaches centered around computing and automated systems. Indeed the twenty-first century witnessed the birth of social networking platforms, which allowed people to communicate virtually regardless of time zone, location, ethnicity, gender, etc. The new environment is associated with several concerns connected to the freedom in communication and assumed trust which is sometimes misused leading to unpleasant cases that should be identified and treated wisely. The chapters included in this book address a number of serious issues associated with social networks. The book is organized in ten chapters. The first chapter is a review of frauds and misuse activities in online social networks. The authors focus on identity manipulation and diffusion of misinformation aspects. The second chapter analyzes the behavior of coordinating deviant groups who created a lot of cyber propaganda against NATO in twitter and blogs. The authors mainly use social network analysis, cyber forensics tools and techniques to uncover the relation between the groups to discover similar groups. The third chapter explains Facebook's ticker and makes a discovery on the privacy, based on the argument of Panopticon or transparency society. It suggests how to protect our privacy to avoid leakage of information to the public unwittingly. The fourth chapter proposes a centrality metric with probabilistic edges for range-limited neighborhoods that is specific to detecting real-time influential spreaders. The fifth chapter sheds light on the benefits of decentralized online social networks and studies its architectures in terms of performance and security. The sixth chapter discusses the future of social networks from different perspectives. The seventh chapter proposes a centrality measure or method for detecting critical nodes especially in directed networks. Strong and weak middle nodes are identified in the network. The eighth chapter considers the issues of investigating criminals on social networks and privacy protection in the context of American legal system. It tries to point out that the proper use of privacy settings may prevent the issues of privacy violations. This chapter tries to explain an expectation gap among legal system, social networking sites, and users from the perspective of privacy risk. The ninth chapter presents a multi-agent-based system with data mining approaches to provide real-time emergency management and analysis by utilizing microblog data streams. The last chapter analyzes Paris attacks and draws lessons which may help to avoid future similar disasters.

This book would not have been put together without the great effort of the authors and reviewers whom we would like to warmly thank for their valuable effort. Reviewers' comments were very constructive to uplift the quality of the content. Springer also deserves a special mention here for their encouragement.

Ankara, Turkey Tansel Özyer
Rourkela, Odisha, India Sambit Bakshi
Calgary, AB, Canada Reda Alhajj

Contents

Frauds in Online Social Networks: A Review

Manoj Apte, Girish Keshav Palshikar, and Sriram Baskaran

Abstract With the widespread use of computers, communications infrastructure, and the Internet, *online social networks* (OSN) have gained a huge popularity in recent years. Unfortunately, the very nature and popularity of OSN have brought about its own share of frauds and misuse. Frauds in OSN refer to activities that result in harassment, loss of money, loss of reputation of a person or an entity, loss of trust in the system or an individual, etc. Due to the complex structure, and information flow in OSN, as well as the relative anonymity of the identity, detection, control and prevention of frauds in OSN is difficult, time-consuming, error-prone and demands an unusually high level of technical finesse from the investigators. In this paper, we begin with a simple typology of OSN frauds and then follow up by describing in detail the nature of each fraud and by reviewing some of the state-of-the-art research done so far (mostly in machine learning, data mining, and text mining) to detect them. Where possible, we stress on the scale and impact of these frauds. We identify manipulation of identities and diffusion of misinformation as two important aspects in the *modus operandi* of most types of OSN frauds. We identify manipulation of identities and diffusion of misinformation as two important aspects in the *modus operandi* of most types of OSN frauds.

1 Introduction

With the widespread use of computers, communications infrastructure, and the Internet, *online social networks* (OSN) have gained a huge popularity in recent years. The uses of OSN range from connecting with friends, creating communities of people having shared interests, creating and sharing content such as articles,

---

M. Apte (✉) · G. K. Palshikar
TCS Research, Tata Consultancy Services Limited, Pune, India
e-mail: manoj.apte@tcs.com; gk.palshikar@tcs.com

S. Baskaran
University of Southern California, Los Angeles, CA, USA
e-mail: sbaskara@usc.edu

T. Özyer et al. (eds.), *Social Networks and Surveillance for Society*, Lecture Notes in Social Networks, https://doi.org/10.1007/978-3-319-78256-0_1

photos, videos, and news, to e-commerce and even to manage political movements and helping during disasters. OSNs have fueled an unparalleled wave of democratization of the web and have offered a voice to diverse peoples and causes. Unfortunately, the very nature and popularity of OSN have brought about its own share of frauds and misuse.

Frauds in OSN refer to activities that result in harassment, loss of money, loss of reputation of a person or an entity, loss of trust in the system or an individual, etc. Frauds in OSN are a subset of the so-called *white collar crimes* or *cybercrimes*. While monetary gain is a major motivation behind most OSN frauds, that is by no means the only one; revenge, prejudices or hatred, for instance, can play a part. Criminal minds now have a new medium to come up with new types of crimes or to devise new methods for older kinds of crimes. The relative anonymity offered by the OSN, the low level of technical abilities among the OSN users (e.g., to configure appropriate security and privacy controls), lack of well-publicized authorities to report to, the complicated structure of OSN, the high-speed and complex information diffusion process, and the transient nature of OSN contents all make it difficult for users as well as investigators to deal with OSN frauds. Hence, detection, control, and prevention of frauds in OSN is difficult, time-consuming, error-prone and demands an unusually high level of technical finesse from the investigators. For these reasons, researchers have been taking a keen interest in devising techniques that can help in detecting OSN frauds.

Governments and lawmakers have realized the scale and impact of OSN frauds, and have been creating legislations to counter them. For instance, online impersonations can lead to civil and criminal liability charges under the CA SB 1411 legislation. Similarly, the Anti-Phishing Consumer Protection Act of 2008 and Utah E-Commerce Integrity act are intended to prohibit phishing for personal gains. There are similar legislations across various countries.

In this paper, we begin with a simple typology of OSN frauds and then follow up by describing in detail the nature of each fraud and by reviewing some of the state-of-the-art research done so far (mostly in machine learning, data mining, and text mining) to detect them. We view an OSN as a network of nodes (people) interacting and sharing information. The information that is shared across the OSN is called the *online social media* or just *social media*. The frauds surveyed here target the OSN users by taking advantage of the specific structural elements of the OSN, the processes of sharing and diffusing information in OSN, and the patterns in the actions of OSN users. Where possible, we stress on the scale and impact of these frauds. We identify manipulation of identities and diffusion of misinformation as two important aspects in the *modus operandi* of most types of OSN frauds. Manipulation of identities is done in identity theft, fake accounts, sybils and cyberbullying or cyberstalking whereas diffusion of misinformation is used in rumours, phishing, brand jacking, and hashtag jacking.

2 Frauds and Its Typologies

We list different types of frauds in OSN based on their purpose and methods.

1. **Type of Fraud:** *Rumours*. **Purpose:** Spread of misinformation. **Method:** Identify a targeted set of users or a community, which are likely to spread the information. Diffuse the information into the identified network [40]. **Outcome:** Cause of panic and instability in OSN and in society at large.
2. **Type of Fraud:** *Identity theft*. **Purpose:** Steal the identity of a user in the same OSN or in another OSN where the victim has not yet registered. **Method:** Identify the victim (one who is operating with unsafe privacy and security configurations), gather all his/her personal information, create a fake profile in the original name with all the stolen details, connect with the friends of the victim and spread the attack [6]. **Outcome:** Breach of trust among the victim's contacts; stealing of personal and confidential information.
3. **Type of Fraud:** *Phishing*. **Purpose:** Stealing and misuse of personal and confidential information. **Method:** Create a replica of a popular website (e.g., the login screen of a bank's website), where the phishing website's name is very similar to the original name. Lead the users towards such websites from untrusted links put up posts in social media and forwards. Ignorant users tend to enter the information which leads to the sharing of personal information [21]. **Outcome:** Stealing of personal information or security credentials; loss of money.
4. **Type of Fraud:** *Brand jacking*. **Purpose:** Mislead or confuse the consumers of a brand. **Method:** Create a fake OSN profile or account, with very similar look and feel to the original. Attract visitors and spread misinformation about the brand to them [2]. **Outcome:** Consumers end up relying on the false information disseminated by the imposter. This false information can be regarding recent events, upcoming products, alleged promotions, and career opportunities. Another outcome could be that the fake profile steals the traffic related to the brand.
5. **Type of Fraud:** *Hashtag jacking*. **Purpose:** Seek attention, attack a particular brand, post unrelated or abusive content, spamming on popular hashtags. **Method:**Hash jacking involves posting repeatedly on trending topics for grabbing attention; or posting links with unrelated tweets; or attacking the business or brand representing the hashtag, using negative or sarcastic comments. Jain et al. [22] **Outcome:** People following the hashtag are exposed to spamming. If the hashtag is related to some disaster, then this might result in panic or misinformation. Brands or businesses represented by the hashtag end up with damaged reputation and brand value.
6. **Type of Fraud:** *Fake Accounts or Sybils*. **Purpose:** Create multiple identities in OSN to gain undue advantage, and launch attacks related to phishing, spamming, stalking, spread of misinformation, etc. [49]. **Method:** Create many false profiles (*sybils*) controlled by a single real person, connect using these fake profiles with as many genuine users as possible, and then overhear communications and act maliciously. **Outcome:** Reduce the reputation of the OSN, gain undue advantage

or influence in the functioning of OSN, launch attacks to gain monetary benefits, manipulate the actions of the OSN.

7. **Type of Fraud:** *Fake Reviews*. **Purpose:** Purposefully boost or harm a product's rating to affect potential user's decision. **Method:** Create multiple duplicate reviews or provide inaccurate representation of an item's quality. Lappas [28] **Outcome:** Impact a potential user's decision due to the boosted or harmed reputation of the item with the help of multiple inaccurate reviews which are purposefully inserted.
8. **Type of Fraud:** *Cyber bullying or cyber stalking*. **Purpose:** To harass a vulnerable person with repeatedly imposed unwanted communication or contact. **Method:** Monitor the target anonymously by gaining entry into the trusted network, by stealing the identity of somebody in the trusted network of the target, often using false identities. Intimidate the target by sending offensive or threatening messages anonymously using fake identities. Dreßing et al. [13] **Outcome:** Harass, intimidate, or blackmail the target.

3 Rumours and Misinformation

Misinformation refers to any false or inaccurate information, often created and shared deliberately to deceive people or an entity. This fraud is based on a simple observation about online behaviour: OSN users sometimes share the received information further without due scrutiny. Diffusion of misinformation is at the root of many types of frauds, such as rumours and spamming, which makes it critical to identify such information, identify its sources, contain its spread, and control the damages. Spread of misinformation can have a high impact in business and politics of a country. Examples of misinformation include fake reviews and political trolling.

A *rumour* is a special case of misinformation, where unverifiable purported "facts" about an event, organization, or person are created deliberately and which spread rapidly across any network of people. Examples of rumours: rioters are going to attack Birmingham children's hospital [32], a volcano has become active after an earthquake in Chile [8], animals have escaped from zoo, etc. Rumour detection in social media is important because of the potential of damage and discontent they may cause as well as due to their unmoderated and rapid spread on social networks. Rumours tend to be about important event or entities, and often give information that is unclear or difficult to verify. Crisis situations often see emergence and spread of rumours in social media, because of the unrest in society and restrictions that prevent the dissemination of truth. The life cycle of a rumour starts as a statement whose veracity is unknown. With no centralized control or moderation in OSN, sharing of information is very easy. So the rumour may start spreading rapidly through the OSN, creating misconceptions with the consumers of such information. A rumour dies down either when the related truth also spreads or when the process of its spread stops. Rumours are more theoretically defined by the *truth value* and *rate of spread*.

Detection of rumours in OSN has become an active area of research. The work on rumour detection can be classified based on the techniques used, viz., (1) information present in the post, (2) spread and propagation features like common substructure of the network promoting or spreading the rumour, and (3) enquiry based detection of rumours. Many of the methods use features from other methods to improve their respective efficiencies. Most of these methods are post facto analysis of the posts that are shared and promoted in the OSN, to build a model that can generalize in detecting new rumours. We now look each class of these methods in more detail.

Information Present in the Post

A natural way to detect a rumour is to analyse contents of each *post* and check whether it contains information that is not credible, not supported by evidence or generally, is of very low quality. If a post is assessed as containing a very low quality information, then it can be tagged as a potential rumour. Castillo et al. [8] concentrated on automatic assessment of credibility of a given set of tweets, as a two-step process. It identifies those topics that are newsworthy and reports a credibility score for each of it. A labeled set of training data is obtained using crowdsourcing on Mechanical Turk for labeling both newsworthy assessment and credibility assessment. A main contribution of this work is that they provide a set of 68 features related to the user (e.g., age, #follower, #friends), the tweet (e.g., length, has URL, has exclamation, has question mark, sentiment score, has emoticons, has mentions of users), the network and the propagation of a set of tweets (e.g., size of the propagation tree related to retweets). Using this exhaustive list of features (see Table 1), they report a high accuracy on detecting low credibility tweets.

Zhang et al. [51] extended this work by stating that shallow features provided by Castillo et al. [8] may not be sufficient to distinguish between rumour and non-rumour. They added features like popularity orientation (similarity of a tweet with any of the current *hot topics*), opinion comments (fraction of comment supporting the tweet), internal and external consistency (similarity between the tweet and contents of any included URL), sentiment polarity and match of the current tweet with the user's historical topics and show that this improves the accuracy of rumour detection process. Chang et al. [9] identified additional features, such as ratio of nouns to verbs and number of quantifiers present, that can be used to detect posts containing low credibility information. Tools like TwitterMonitor are typically used to identify *topics* that the posts contain, where a topic is characterized by a set of compulsory words and a set of optional words.

Table 1 Some features for detecting rumours using information in posts

Type	Feature
Content	Number_of_@, Number_of_#, Number_of_sentiment_words, Geo_location, Number_of _URLs_present, Number_of_words, Client_type, Time_span
User	User_type, Number_of_followers, Number_of_followees, Has_description, Gender, User_avatar_type, Registration_time, User_name_type
Propagation	Number_of_comments, Number_of_reposts, Is_repost
Topic	Average_registration_age, Average_status_count, Fraction_of_URLs, Average_number_of_followers, Fraction_of_authors_who _are_verified, Fraction_of_authors_having_description, Fraction_of_@, Fraction_of_#
Multimedia	Has_multimedia, Time_span

Spread and Propagation

Spread (i.e., diffusion) of rumours has an appealing similarity with spread of diseases, and hence epidemiological models have been adapted for modeling how rumours spread in OSN. Most epidemiological models of disease spread are stochastic processes, typically Markov chains, where the states correspond to individuals who are *susceptible* (S), *infected* (I), or *recovered* (R) and transitions occur between them over time. The other class of these models uses linked differential equations. One goal of these models is to predict the temporal and spatial extent of the disease spread. In case of rumours, a user is susceptible if the rumour has not yet reached him/her, infected if the rumour reaches him/her and he/she believes it and propagates it further, and recovered if he/she was previously infected but has now stopped propagating it further.

Jin et al. [23] proposed a differential equations based model for diffusion of rumours in OSN. They dropped the state recovered, but added two new states called *exposed* (E) where a user has received a rumour but has taken some time (*incubation*) before he/she started propagating, and *skeptic* (Z), where a user has received a rumour but does not propagate it. They use this model to derive a quantity called R_{SI}, which they compute for any candidate rumour and any candidate having a low value is potentially a rumour:

$$R_{SI} = \frac{(1-p)\beta + (1-l)b}{\rho + \epsilon} \tag{1}$$

where β is the $S - I$ contact rate, b is the $S - Z$ contact rate, ρ is the $E - I$ contact rate, ϵ is the incubation rate, l is the $S \rightarrow Z$ probability given contact with Z, and p is the $S \rightarrow I$ probability given contact with I or E. They demonstrated utility of this model to detect true news (e.g., Boston marathon bombing and Pope resignation) from rumours (e.g., Castro's death and Obama injury).

People react to the tweets (or any post) that they receive, typically ignoring, supporting or denying the content. This suggests an important way to detect a rumour by looking at how many tweets supported or denied it. Lukasik et al. [32] uses Gaussian processes to detect rumours using such features.

Wu et al. [44] argued that the flat structure of features extracted from the tweet and the associated topic is not enough to detect rumours. They added features related to the propagation and spread of the tweet (e.g., average doubt, average surprise, average emoticon, repost time score). These features are computed using a *propagation tree* built for every message, with the initial message at the root and all the reposting and replies as children. Every message is associated with the user who posted it, the timestamp, the client information, and the media if any associated with it. The SVM classifier uses a hybrid novel random walk graph kernel and normal radial basis function (RBF). The similarity between the two graphs is calculated using the random walk kernel and the high dimensional distance between two vectors of traditional and semantic features is calculated by RBF kernel.

One important aspect of rumour spreading is to convince influential users to support and spread the rumour. Hence it is important for the rumour mongers to identify such influential users. There are several criteria to measure the level of influence of a user in an OSN, such as those based on centrality or betweenness. A popular criterion is $\mu - PCI$, which is defined as follows. The $\mu - PCI$ of a node v is equal to k, if there are at most $\mu \cdot k$ nodes in the μ-hop neighbourhood of v with degree $\geq k$, and the rest of the nodes in that neighborhood have a degree $\leq k$. Wang et al. [40] present another measure for identifying influential users called Fractional and Directed Power Community Index (FDPCI). They define a weight W_{ij} between users i and j as:

$$W_{ij} = \frac{n_c}{n_t} + \frac{t_d}{t_p} + \frac{f_s}{f_t} + v \tag{2}$$

where n_c is the number of shared contacts between i and j, n_t is the total number of contacts of i, t_d is the number of active days among a total period of t_d days, f_s is the number of shared factors among a total of f_t factors (like gender and location), and v is a Boolean value indicating whether user i is verified or not. Weight of a node i is the sum of the weights of all its outgoing edges. FDPCI of a node i is the same as $\mu - PCI$ except that they set $\mu = 1$ and use the weight of a node instead of its degree.

4 Brand Jacking and Hashtag Jacking

On Wikipedia, a *hashtag* is defined as a type of label or metadata tag used on social network and microblogging services, which makes it easier for users to find messages with a specific topic, event, person, entity, theme, or content. Any combination of characters preceded by a hash symbol creates a hashtag. Users who contribute to the topic of interest use the hashtag in their tweets and interested users follow the specified hashtags. This arrangement is beneficial for discussions on these topics. If used by enough users the hashtag can trend and attract more users to the specific discussion.

Popularity of a hashtag attracts spammers, who hijack the hashtag either for personal gain or for confusing the followers of the hashtag. Spammers work to change the profile of the hashtag resulting in followers getting unwanted and false information. In Twitter domain this is referred to as *hashtag jacking*. For example, a hashtag like `#ObamaCare` is an attractive target for both supporters and opponents of the associated healthcare program in the USA. The Republicans may use it to criticize the program and the Democrats may use it to praise the program. An article in the Forbes Magazine[1] gives several examples of hijacking of hashtags. For example, McDonald's started the hashtag `#McDStories` to collect heart-warming stories about Happy Meals; instead, it was hijacked and got linked to another hashtag `#McDHorrorStories` having tweets deriding McDonald's.

There are two main characteristics of hashtag jacking: first there is a sudden burst of tweets using it and second, there is a change in the nature of the information polarity or sentiment associated with the hashtag. Most techniques for detection of hashtag hijacking make use of these characteristics.

Hadgu et al. [19] studied the polarization of such political hashtags on Twitter over time. They used a simple rule-based method to identify peaks in the number of tweets using the hashtag over time, which they called the *change points*. Then they used a simple algorithm to identify the current *leaning* (Left or Right) of the tweets associated with the hashtag up to a time instant, showing that these change points or jumps in leanings happen due to hashtag hijackers.

Sedhai and Sun [37] have released a dataset of 14 million tweets matching certain hashtags and then conducted systematic annotation of the tweets being spam (e.g., tweets containing adult content or offers of quick money or free gifts, etc.) and ham (i.e., non-spam). They grouped highly similar tweets into clusters, labeled the clusters and then used an EM-based algorithm to predict labels for unlabeled tweets. They defined a *spammy index* $si(t)$ of a hashtag t to quantify the extent a hashtag is being used in spam tweets, pointing to the possibility of hashtag being hijacked by spammers:

$$si(t) = \log_2(df(t, D)) \times \frac{df(t, D_s)}{df(t, D)} \tag{3}$$

Here, $df(t, D_s)$ denotes the document frequency or the number of tweets containing hashtag t in the tweets collection D, $D_s \subset D$ is the set of spam tweets. A hashtag is more spammy i.e., potentially hijacked if either its document frequency is high or the probability of its being used in the spam is high. Then they define a *spammy index odds ratio* (OR) for words in tweets and show that tweets containing

[1]http://www.forbes.com/sites/kashmirhill/2012/01/24/mcdstories-when-a-hashtag-becomes-a-bashtag/.

hashtags having high spammy indexes also tend to contain words with high OR values.

Lamba et al. [27] contains study on *firestorms*, which are sudden outbursts of negative attention to a person or an organization, typically in cases of controversy or outrage. They found and statistically analysed 80 examples of firestorms. They found that the average peak volume of a firestorm on Twitter is about 50,000 (in 2011). Firestorms are likely to be related to hashtag hijacking. They mentioned that the hashtag `#myNYPD`, which was originally started by the New York Police Department to collect positive stories, was later hijacked to highlight police brutalities.

Jain et al. [22] proposes a tool HashJacker which detects and analyzes hijacking of Hashtag. Hashjacker calculates TFIDF score for top words in tweets related to one hashtag, does matching with a tweet, and reports low scoring tweets as most unrelated tweets for that specific hashtag. HashJacker can be used to detect spam and junk distributed through popular trending hashtags. Once these hashtags are identified as spammed or hijacked, tracking the irrelevant tweets can lead to the users who are spamming these popular hashtags.

5 Phishing

Direct attacks on people can be carried out in different ways. *Phishing* is one way to acquire sensitive information by luring and deceiving the recipient. The recipient inadvertently shares the information by considering the fraudster as a trustworthy entity in the OSN. Typically, phishing results in identity theft, denial of access email, and financial loss. Phishing is quite widespread in OSN. The Anti-Phishing Work Group (http://www.antiphishing.org) detected a total of 27,221 unique phishing URLs in January 2007 [18]. In 2010, 43% of all the OSN users were targets of phishing attacks and in 2012, around 20% of all phishing attacks targeted FaceBook [3]. According to Gartner, phishers stole $1.7 billion in 2008 [42].

Jagatic et al. [21] describes the phishing methods in OSN, the intended targets and the ease by which phishing can be carried out in social media. A typical method for phishing consists of the following steps.

1. The user is provided with a topic of his/her interest or a currently popular topic, along with a URL (typically shortened using bit.ly and tinyurl.com), drawing the user to click on to it.
2. When the user clicks the URL, it leads to a website that deliberately looks similar (*clone*) to one of the trusted sites.
3. The user enters personal information including passwords.
4. The application captures the information and goes to an error page to mislead the user that there was some error in the system. The user is unaware of that their personal information has been stolen.

One way to subvert phishing is to educate and train the users. Another way is to use a black-list of known phishing websites, where it becomes crucial to have an up-to-date blacklist. Finally, one can develop automated tools that detect either a phishing email or a phishing website. Khonji et al. [26] is a comprehensive survey of tools and techniques for detection of phishing. We will survey only a few papers related to phishing.

Fette et al. [17] is applied classification techniques to detect phishing emails. They used features such as IP-based URL (e.g., http://192.168.0.1/paypal.cgi?fixaccount), age of the domain name, non-matching of displayed URL and actual URL, presence of one malicious link among many other official links in the email, whether email is HTML or plain-text, number of links in the email, number of domains in all the links in the email, number of dots in the link, contains javascript, etc.

Garera et al. [18] applied classification techniques to detect phishing URLs. In addition to the features mentioned above, they also used features like: use of a valid-looking domain names but which are obfuscating, too long, misspelt or are redirected to another website, presence of keywords like small `login, banking, secure, account`, and so forth.

Abdelhamid et al. [1] applies association rule based classification to the problem of detection of phishing URLs. They use similar features as mentioned earlier, and some new features like use of pop-up window, web-site traffic, etc. Wenyin et al. [41] proposed the use of image similarity (broken into block-level, layout and style similarity) to detect similar-looking (em cloned) websites, which are commonly used in phishing.

Phishing in Twitter is harder to detect because of real-time spread, short size of tweets, and use of shortened URLs. Aggarwal et al. [3] uses classification techniques (Naive Bayes, Decision Trees, and Random Forest) to detect whether a tweet containing a URL is phishing or not. In addition to the features mentioned earlier, they use novel features such as number of redirections, Levenshtein distance between original and final landing URL, number of @, number of hashtags, presence of trending hashtags, length of tweet, age of Twitter account, number of followers, number of followees, etc.

To help combat phishing, Google publishes a blacklist of phishing URLs and phishing URL patterns. Whittaker et al. [42], Ma et al. [33], Ludl et al. [31] have tried to automate the process of identifying phishing websites given a list of blacklisted websites (typically obtained from sources like phishtank.com, or SpamScatter).

Many surveys and studies have been conducted to detect and mitigate phishing attacks. There are numerous tools (typically web browser plug-ins) developed to detect phishing websites. Zhang et al. [50] and Egelman et al. [15] carried out empirical studies of the different phishing website identification tools and their efficiency in detecting them. They tested the tools manually as well as in an automated manner.

6 Identity Theft

Federal Trade Commission defines *identity theft* as knowingly transferring or using, by unlawful means, the identification of another person with intent to commit, aid, or abet any unlawful activity.[2] The stolen identity information (e.g., personal details (like DOB, phone number), passwords, SSN, credit card details, bank account details, etc.) is often used to cause financial loss to the original user. Identity theft is widespread; e.g., 9.3 million adults had been victimized by some form of identity theft in 2004 [36]. In the USA, estimated losses due to identity theft have been about $52.6 billion p.a. in the period 2003–2006 [34]. The number of identity theft victims in 2013 was about 13 million, and economic losses were $18 billion [45].

Despite identity theft being widespread, the scope of the problem, it extent, and its importance are not yet well understood [20]. Allison et al. [5] offers a criminological analysis of a group of identity theft cases in Florida (e.g., age, race, gender, employment status, etc. for both victims and offenders).

It is also difficult to determine such frauds in the online social media due to the following reasons:

1. The victims are unaware that their identity has been stolen and/or misused for some unlawful purpose.
2. Even if users come to know about it, they fail to report it to the relevant authorities. Also, in case of social media, the authorities are not well known.

In general, the goals of identity thefts are: to cause financial loss to the victim, to tarnish the reputation of a person in social media, to gain undue benefits and services, to participate in crimes under the guise of another person, and to evade detection by law enforcement [4]. Identity theft is carried out in two ways [6]: (1) *profile cloning*: the thief copies the details from the identity (profile) of an existing user and then sends requests to friends/followers/contacts of the victim; or (2) *cross-site profile cloning*: the thief identifies an existing user and creates an identity in platforms where the user doesn't have an account and then sends requests to the victim's friends in the other network.

The weak link in this modus operandi obviously is the user. Many aren't aware of the security measures to comply with and show much implicit trust. It is critical to educate the users with such information which creates a first line of defence. Developers of social networking websites have taken care to restrict the use of bots and engines that automatically crawl their pages for personal and private information. CAPTCHAs are commonly used to counter automated crawling. FaceBook uses an upgraded system of reCAPTCHA [39] to reduce access by automated crawlers. Still, many social networking sites don't take enough care to avoid such attacks.

---

[2]http://www.ftc.gov/sites/default/files/itadact.pdf/.

Bilge et al. [6] report experiments where they create fictitious (fake) profiles or use a crawler to clone profiles, send friend requests to the contacts of the stolen identities and also send fake messages using the stolen or fictitious identities. They found significantly high acceptance rates for contact requests and significant high click through rate for messages sent using stolen or fictitious profiles.

Wu et al. [43] consider in situ identity theft in FaceBook, where a legitimate user's device is accessed by somebody else; e.g., due to saved passwords, stolen devices, use of cookies, low protection for physical access (e.g., no PIN), etc. They use features derived from the stalker's browser actions (e.g., click like, add comment, delete comment, click hyperlinks, view messages, view photos etc.) and use SVM to detect identity theft. They reported 80% detection accuracy within 2 min.

Egele et al. [14] used a combination of anomaly detection and classification to detect compromised accounts in FaceBook and Twitter, using features derived from user actions. First, they assigned an anomaly score to each message sent by the user, using features such as language, time of day, whether a third party application was used to send the message, message topic, links in message, user mentions (e.g., using @ in Twitter), etc. They used comparison with the typical behavioural profile of the user to compute this anomaly score. They also built a classification model for detecting compromised accounts. They report an account as compromised if it has a high anomaly score and it is classified as suspicious by the classifier.

Yang et al. [45] collected 3500 news stories over a 45-day period reporting identity theft, used text mining to extract information such as target user, type of information stolen, location, loss amount, and performed several analyses of this data.

7 Cyber Stalking/Bullying

Stalking is repeated unwanted contact or harassment over a significant time period (e.g., for at least 2 weeks) by an individual or a group towards another person. When this pursuit is carried out using electronic devices it becomes *cyberstalking*. Cyberstalking is done through various electronic means, such as emails, instant messengers, chatrooms, message boards, and OSN such as FaceBook or Twitter. Methods of cyberstalking include repeated contact, sending insulting, provocative or hateful messages, undue aggression, use of extremely offensive, violent or explicit language, spreading of falsehoods, false accusations, damage to reputation, publishing forged images, publishing private information either on pornographic websites or elsewhere, luring the victim for a physical meeting or for sending money, etc. Women, children, students, and celebrities are often subjected to cyberstalking. Cyberstalking can lead to mental anxiety, panic, depression, sleep problems, fear, and even nervous breakdowns among the victims. Cyberstalkers are typically male, better educated and being technically more adept than their victims, they take advantage of the relative anonymity of the online platforms. It is often difficult for victims to prevent or stop acts of cyberstalking on their own. *Cyberbullying* is similar to cyberstalking and involves the additional step of some aggressive and violent act carried out by an individual or a group.

Cyberstalking is fairly widespread. Dreßing et al. [13] conducted an online survey of 6379 participants of German OSN StudiVZ where they found 6.3% prevalence of cyberstalking (399 victims). They found that among the victims, 80.5% were females, 54% were single and the average age of victims was 24.6 years. They also showed that victims of cyberstalking had a poorer score in terms of WHO-5 Well-being Index. They showed that the perpetrator is often known to the victim: typically, a friend, ex-partner, acquaintance, or a relative. They also debated when cyberstalking can be called as a crime; e.g., whether keeping track of someone's FaceBook profile can be properly called cyberstalking.

Burmester et al. [7] created a cyberstalking threat model including the stalker's profile, law enforcement constraints and then suggested a cryptographic approach for the tracking of cyberstalkers. They proposed the use of a transparent and secure Monitor module (installed on every potential victim's machine) to "eavesdrop" on any messages exchanged between a victim and perpetrator (i.e., the trail) and collect data and forensic evidence for cyberstalking, if it has happened.

Since the textual messages involved in cyberstalking are of a specific nature (as discussed above), Dinakar et al. [12] attempted to detect cyberbullying by building classifiers to detect such messages. They downloaded about 50,000 textual comments posted on YouTube videos, clustered them into groups related to *sexuality* (sexists attacks on women, attacks on sexual minorities), *race or culture* (racial or cultural stereotypes, mocking of cultural practices or traditions), *intelligence* (attacks on intelligence or mental capabilities) or none. They manually assigned one of these 4 class labels to 1500 comments from each cluster. Finally they used various classification techniques on this training data to predict class labels for unseen comments. While the classifiers were reasonably accurate, not surprisingly, they failed on sarcastic comments.

Since most cyberstalkers are male, and also men and women have somewhat distinguishable writing styles, Dadvar et al. [10] built gender specific classifiers to detect harassment kind of messages. They used 381,000 posts from MySpace OSN, of which 34% were written by female and 64% were written by male authors. They manually labeled 2200 posts as *harassing* or *not-harassing*. They used four features: profane words, second person pronouns, other personal pronouns, and the sum of TFIDF weights of all the words in each sentence. Then they used SVM to build a classifier to the class labels, and showed that the use of gender-specific features improves the prediction accuracy. Yin et al. [47] is similar, but they used sentiment features, and not gender-specific features.

8 Fake Reviews and Fake Likes

Online reviews are a valuable source of information potential customers of popular brands, products, and e-commerce companies. Online reviews provide valuable information and aid in decision-making process of customers. Writing fake reviews is a form of attack that purposefully harms or boosts a product or organization (e.g., a hotel) and its reputation [28]. Detecting fake reviews is difficult even for humans, and hence there is a lot of work on automated fake review detection. We review only some of the work here.

A natural approach to detect fake reviews is to examine the contents of the reviews. For example, if a review is a duplicate (i.e., there are many other reviews very similar to this one), then it is likely to be fake [24]. Liu et al. [30] examined whether a review was *useful* or *helpful*—fake reviews are unlikely to be helpful. Another prominent approach for fake review detection is to examine the behaviour of the reviewers. Jindal et al. [25] represented each review as records, with features about reviewer, brand, product, etc. and the rating given by the reviewer as the class label. Using association mining, they identified rules that can detect reviews having *unexpected* support, confidence, attribute distribution, or attribute value. For example, a rule might detect that a reviewer wrote only positive reviews for one brand and only negative reviews for another brand. Lim et al. [29] attempted to detect spamming reviewers by checking whether a reviewer is targeting a specific product or whether his *rating behaviour* differs significantly from other reviewers.

Lappas [28] proposed two different aspects of the fake reviews namely, authenticity (authentic-looking and convincing) and impact (on target's reputation). Authenticity was modeled in terms of stealth, coherence, and readability. They defined models for evaluating these aspects and used them to detect fake reviews. Mukherjee et al. [35] targeted the scenario where the reviewers were colluding on what is to be written. They created different groups of reviewers using frequent itemset mining and built different indicators like Group Time Window, Group Deviation, Group Content Similarity, and Group Early Time Frame. Feng et al. [16] postulated that distributional footprint of reviews for a product would be different with the presence of fake reviews. Amazon and TripAdvisor reviews are considered for this analysis and models for different types of review comments are provided.

9 Fake Accounts/Sybils

Due to lack of any serious identity verification, it is easy for a single real person to create many fake or anonymous accounts (*sybils*) with false names on various social media platforms. Fake accounts are widespread; according to CNN, there were 83 million fake accounts in FaceBook in 2012.[3] According to FaceBook's own estimate, between 87 to 137 million fake accounts existed in 2013.[4] While some fake accounts are relatively benign, others often carry out cyber-crimes such as stalking, creating fake product reviews, manipulating responses to online advertisements (*click frauds*), spreading false information, creating false impressions (e.g., "likes" or followers), trolling, phishing, spamming and spreading malware, often colluding with each other during these attacks. In this section, we focus on fake accounts in OSN platforms, like FaceBook or Twitter. Fake accounts are created with properly filled up profiles containing realistic but false information and photographs, which makes their detection a challenge. OSN platforms have simple detection rules such as (1) presence of blacklisted URLs, suspect keywords or extremely abusive language in the content, or (2) attempts to create large number of accounts from a single IP address within a relatively short time window. Such rules and additional schemes (e.g., protection through CAPTCHAs or allowing users to report spam or abusive accounts) are easy to bypass.

One characteristic of fake accounts is that they are often created in a group, and are linked to each other, thus forming a community. Hence community detection approaches have been proposed to detect fake accounts. Yu et al. [48] assume that the OSN topology is such that sybils and honest users each form a community of their own, with many *sybil edges* (between sybil nodes), many *honest edges* (between honest nodes) and relatively few *attack edges* crossing over. They propose a simple decentralized scheme where an honest node H accepts a friend request from a suspect node S if two random walks of length w each—one starting at H and another starting at S—intersect. This is based on the facts that a random walk from an average honest user is highly likely to stay within the honest community; and that two random walks from two honest nodes are highly likely to intersect within w steps. Algorithms SybilLimit [49], and SybilInfer [11] are similar in spirit.

A key initial activity for a sybil is to acquire many friends, typically by proactively sending unsolicited friend requests to normal users. Yang et al. [46] report that only 26% of friend requests sent out by sybils are accepted, as compared to 79% for genuine users. Sybils also send out a large number of friend requests per time interval (e.g., per hour). Using features such as invitation frequency, incoming and outgoing requests accepted, and clustering coefficient of the friend graph of an account, they created a training dataset from nearly 560,000 RenRen accounts that were detected and banned using rule-based techniques earlier. They trained an SVM classifier on this data and used it predict whether an account in RenRen (a Chinese OSN similar to FaceBook) is sybil or not. From August 2010 to February 2011, RenRen used this system to detect and subsequently ban nearly 100,000 Sybil accounts. They also showed that sybil accounts have significantly different click behaviour: 63.3% for friending (12.3% for normal users), 4% for photos and

[3]http://edition.cnn.com/2012/08/02/tech/social-media/facebook-fake-accounts/.

[4]https://thenextweb.com/facebook/2014/02/03/facebook-estimates-5-5-11-2-accounts-fake/.

messages (68% for normal users), 27.3% for visiting profiles (9.6% for normal users), etc.

Many web-sites (e.g., YouTube, FaceBook) allow users to vote on the contents, such as news, photos, and products. Sybils can pollute such systems by outvoting genuine users, thus promoting low-quality content. SumUp [38] counters this by proposing a centralized vote collection mechanism that ensures that most votes are from honest users and votes from sybils are rejected with high probability. SumUp dynamically decides and assigns a *capacity* c_e to each edge e, and uses max-flow like algorithm to ensure that only c_e votes are collected. SumUP guarantees with high probability that the number of bogus votes is bounded by the number of attack edges.

10 Conclusions

In this paper, we began with a simple typology of OSN frauds and then followed up by describing in detail the nature of each fraud and by reviewing some of the state-of-the-art research done so far (mostly in machine learning, data mining and text mining) to detect them. While much research has been done to detect OSN frauds, the problems are far from being solved. The intersection of e-commerce, electronic payment systems, and the highly evolving nature of OSNs is a potent mix and a fertile hunting ground for white collar criminals. In the years to come, we will certainly see newer types of frauds and newer methods. There is a need to constantly develop clever methods for detection, prevention, and control of OSN frauds.

References

1. N. Abdelhamid, A. Ayesh, F. Thabtah, Phishing detection based associative classification data mining. Expert Syst. Appl. **41**(13), 5948–5959 (2014)
2. N.A. Adawi, Social networking and "brand-jacking"–is it infringement? (2010)
3. A. Aggarwal, A. Rajadesingan, P. Kumaraguru, Phishari: automatic realtime phishing detection on twitter, in *Proceedings of IEEE eCrime Researchers Summit* (2012)
4. E. Aïmeur, D. Schőnfeld, The ultimate invasion of privacy: identity theft, in *2011 Ninth Annual International Conference on Privacy, Security and Trust (PST)* (IEEE, New York, 2011), pp. 24–31
5. S.F.H. Allison, A.M. Schuck, K.M. Lersch, Exploring the crime of identity theft: prevalence, clearance rates, and victim/offender characteristics. J. Crim. Just. **33**(1), 19–29 (2005)
6. L. Bilge, T. Strufe, D. Balzarotti, E. Kirda, All your contacts are belong to us: automated identity theft attacks on social networks, in *Proceedings of the 18th International Conference on World Wide Web* (ACM, New York, 2009), pp. 551–560
7. M. Burmester, P. Henry, L.S. Kermes, Tracking cyberstalkers: a cryptographic approach. ACM SIGCAS Comput. Soc. **35**(3), 2 (2005)
8. C. Castillo, M. Mendoza, B. Poblete, Information credibility on twitter, in *Proceedings of the 20th International Conference on World Wide Web* (ACM, New York, 2011), pp. 675–684

9. T. Chang, P.Y. Hsu, M.S. Cheng, C.Y. Chung, Y.L. Chung, Detecting fake review with rumor model—case study in hotel review, in *International Conference on Intelligent Science and Big Data Engineering* (Springer, Berlin, 2015), pp. 181–192
10. M. Dadvar, F.M.G. de Jong, R.J.F. Ordelman, R.B. Trieschnigg, Improved cyberbullying detection using gender information (2012)
11. G. Danezis, P. Mittal, Sybilinfer: detecting sybil nodes using social networks, in *NDSS*, San Diego, CA (2009)
12. K. Dinakar, R. Reichart, H. Lieberman, Modeling the detection of textual cyberbullying. Soc. Mob. Web **11**(02), 11–17 (2011)
13. H. Dreßing, J. Bailer, A. Anders, H. Wagner, C. Gallas, Cyberstalking in a large sample of social network users: prevalence, characteristics, and impact upon victims. Cyberpsychol. Behav. Soc. Netw. **17**(2), 61–67 (2014)
14. M. Egele, G. Stringhini, C. Kruegel, G. Vigna, Compa: detecting compromised accounts on social networks, in *Proceedings of Symposium on Network and Distributed System Security (NDSS)* (2013)
15. S. Egelman, L.F. Cranor, J. Hong, You've been warned: an empirical study of the effectiveness of web browser phishing warnings, in *Proceedings of the SIGCHI Conference on Human Factors in Computing Systems* (ACM, New York, 2008), pp. 1065–1074
16. S. Feng, L. Xing, A. Gogar, Y. Choi, Distributional footprints of deceptive product reviews. ICWSM **12**, 98–105 (2012)
17. I. Fette, N. Sadeh, A. Tomasic, Learning to detect phishing emails, in *Proceedings of the International Conference on World Wide Web (WWW2007)* (2007), pp. 55–56
18. S. Garera, N. Provos, M. Chew, A.D. Rubin, A framework for detection and measurement of phishing attacks, in *Proceedings of (WORM 2007)* (2007)
19. A.T. Hadgu, K. Garimella, I. Weber, Political hashtag hijacking in the U.S. (2013)
20. C.J. Hoofnagle, Identity theft: making the known unknowns known. Harv. J. Law Technol. **21**, 97 (2007)
21. T.N. Jagatic, N.A. Johnson, M. Jakobsson, F. Menczer, Social phishing. Commun. ACM **50**(10), 94–100 (2007)
22. N. Jain, P. Agarwal, J. Pruthi, Hashjacker-detection and analysis of hashtag hijacking on twitter. Int. J. Comput. Appl. **114**(19), 17–20 (2015)
23. F. Jin, E. Dougherty, P. Saraf, Y. Cao, N. Ramakrishnan, Epidemiological modeling of news and rumors on twitter, in *Proceedings of the 7th Workshop on Social Network Mining and Analysis* (ACM, New York, 2013), p. 8
24. N. Jindal, B. Liu, Opinion spam and analysis, in *Proceedings of the 2008 International Conference on Web Search and Data Mining* (ACM, New York, 2008), pp. 219–230
25. N. Jindal, B. Liu, E.-P. Lim, Finding unusual review patterns using unexpected rules, in *Proceedings of the 19th ACM International Conference on Information and Knowledge Management* (ACM, New York, 2010), pp. 1549–1552
26. M. Khonji, Y. Iraqi, A. Jones, Phishing detection: a literature survey. IEEE Commun. Surv. Tutorials **15**(4), 2091–2121 (2013)
27. H. Lamba, M.M. Malik, J. Pfeffer, A tempest in a teacup? analyzing firestorms on twitter, in *2015 IEEE/ACM International Conference on Advances in Social Networks Analysis and Mining (ASONAM)* (IEEE, New York, 2015), pp. 17–24
28. T. Lappas, Fake reviews: the malicious perspective, in *International Conference on Application of Natural Language to Information Systems* (Springer, Berlin, 2012), pp. 23–34
29. E.-P. Lim, V.-A. Nguyen, N. Jindal, B. Liu, H.W. Lauw, Detecting product review spammers using rating behaviors, in *Proceedings of the 19th ACM International Conference on Information and Knowledge Management* (ACM, New York, 2010), pp. 939–948
30. J. Liu, Y. Cao, C.-Y. Lin, Y. Huang, M. Zhou, Low-quality product review detection in opinion summarization, in *EMNLP-CoNLL*, vol. 7 (2007), pp. 334–342
31. C. Ludl, S. McAllister, E. Kirda, C. Kruegel, On the effectiveness of techniques to detect phishing sites, in *International Conference on Detection of Intrusions and Malware, and Vulnerability Assessment* (Springer, Berlin, 2007), pp. 20–39

32. M. Lukasik, T. Cohn, K. Bontcheva, Classifying tweet level judgements of rumours in social media, in *Proceedings of the 2015 Conference on Empirical Methods in Natural Language Processing (EMNLP2015)* (2015), pp. 2590–2595
33. J. Ma, L.K. Saul, S. Savage, G.M. Voelker, Beyond blacklists: learning to detect malicious web sites from suspicious urls, in *Proceedings of the 15th ACM SIGKDD International Conference on Knowledge Discovery and Data Mining* (ACM, New York, 2009), pp. 1245–1254
34. R.T. Mercuri, Scoping identity theft. Commun. ACM **49**(5), 17–21 (2006)
35. A. Mukherjee, B. Liu, N. Glance, Spotting fake reviewer groups in consumer reviews, in *Proceedings of the 21st International Conference on World Wide Web* (ACM, New York, 2012), pp. 191–200
36. G.R. Newman, M.M. McNally, Identity theft literature review. US Department of Justice Document No. 210459 (2005)
37. S. Sedhai, A. Sun, Hspam14: a collection of 14 million tweets for hashtag-oriented spam research, in *Proceedings of the 38th International ACM SIGIR Conference on Research and Development in Information Retrieval* (ACM, New York, 2015), pp. 223–232
38. D.N. Tran, B. Min, J. Li, L. Subramanian, Sybil-resilient online content voting, in *NSDI*, vol. 9 (2009), pp. 15–28
39. L. Von Ahn, B. Maurer, C. McMillen, D. Abraham, M. Blum, recaptcha: human-based character recognition via web security measures. Science **321**(5895), 1465–1468 (2008)
40. D. Wang, A. Musaev, C. Pu, Information diffusion analysis of rumor dynamics over a social-interaction based model, in *2016 IEEE 2nd International Conference on Collaboration and Internet Computing (CIC)* (IEEE, New York, 2016), pp. 312–320
41. L. Wenyin, G. Huang, L. Xiaoyue, Z. Min, X. Deng, Detection of phishing webpages based on visual similarity, in *Proceedings of the International Conference on World Wide Web(WWW2005)* (2005), pp. 1060–1061
42. C. Whittaker, B. Ryner, M. Nazif, Large-scale automatic classification of phishing pages, in *NDSS*, vol. 10 (2010)
43. S.-H. Wu, M.-J. Chou, C.-H. Tseng, Y.-J. Lee, K.-T. Chen, Detecting in-situ identity fraud on social network services: a case study with facebook, in *Proceedings of the 23rd International Conference on World Wide Web(WWW2014)* (2014), pp. 401–402
44. K. Wu, S. Yang, K.Q. Zhu, False rumors detection on sina weibo by propagation structures. In *2015 IEEE 31st International Conference on Data Engineering (ICDE)* (IEEE, New York, 2015), pp. 651–662
45. Y. Yang, M. Manoharan, K.S. Barber, Modelling and analysis of identity threat behaviors through text mining of identity theft stories, in *Proceedings of IEEE Joint Intelligence and Security Informatics Conference (JISIC)* (2014), pp. 184–191
46. Z. Yang, C. Wilson, X. Wang, T. Gao, B.Y. Zhao, Y. Dai, Uncovering social network sybils in the wild. ACM Trans. Knowl. Discov. Data **8**(1), 2 (2014)
47. D. Yin, Z. Xue, L. Hong, B.D. Davison, A. Kontostathis, L. Edwards, Detection of harassment on web 2.0, in *Proceedings of the Content Analysis in the WEB 2.0 (CAW2.0) Workshop at WWW2009* (2009)
48. H. Yu, M. Kaminsky, P.B. Gibbons, A. Flaxman, Sybilguard: defending against sybil attacks via social networks, in *ACM SIGCOMM Computer Communication Review*, vol. 36 (ACM, New York, 2006), pp. 267–278
49. H. Yu, P.B. Gibbons, M. Kaminsky, F. Xiao, Sybillimit: a near-optimal social network defense against sybil attacks, in *IEEE Symposium on Security and Privacy, 2008. SP 2008* (IEEE, New York, 2008), pp. 3–17
50. Y. Zhang, S. Egelman, L. Cranor, J. Hong, Phinding phish: evaluating anti-phishing tools (2006)
51. Q. Zhang, S. Zhang, J. Dong, J. Xiong, X. Cheng, Automatic detection of rumor on social network, in *National CCF Conference on Natural Language Processing and Chinese Computing* (Springer, Berlin, 2015), pp. 113–122

Leveraging Social Network Analysis and Cyber Forensics Approaches to Study Cyber Propaganda Campaigns

Samer Al-Khateeb, Muhammad Nihal Hussain, and Nitin Agarwal

Abstract In today's information technology age, our political discourse is shrinking to fit our smartphone screens. Further, with the availability of inexpensive and ubiquitous mass communication tools like social media, disseminating false information and propaganda is both convenient and effective. Groups use social media to coordinate cyber propaganda campaigns in order to achieve strategic and political goals, influence mass thinking, and steer behaviors or perspectives about an event. In this research, we study the online deviant groups (ODGs) who created a lot of cyber propaganda that were projected against the NATO's Trident Juncture Exercise 2015 (TRJE 2015) on both Twitter and blogs. Anti-NATO narratives were observed on social media websites that got stronger as the TRJE 2015 event approached. Call for civil disobedience, planned protests, and direct action against TRJE 2015 propagated on social media websites. We employ computational social network analysis and cyber forensics informed methodologies to study information competitors who seek to take the initiative and the strategic message away from NATO in order to further their own agenda. Through social cyber forensics tools, e.g., Maltego, we extract metadata associated with propaganda-riddled websites. The extracted metadata helps in the collection of social network information (i.e., friends and followers) and communication network information (i.e., network depicting the flow of information such as tweets, retweets, mentions, and hyperlinks). Through computational social network analysis, we identify influential users and powerful groups (or the focal structures) coordinating the cyber propaganda campaigns. The study examines 21 blogs having over 18,000 blog posts dating back to 1997 and over 9000 Twitter users for the period between August 3, 2014, and September 12, 2015. These blogs were identified, crawled, and stored in our database that is accessible through the Blogtrackers tool. Blogtrackers tool further helped us identify the activity patterns of blogs, keyword patterns, and the influence a blog or a blogger has on the community, and analyze the sentiment diffusion in the community.

S. Al-Khateeb (✉) · M. N. Hussain · N. Agarwal
Department of Information Science, University of Arkansas at Little Rock, Little Rock, AR, USA
e-mail: sxalkhateeb@ualr.edu; mnhussain@ualr.edu; nxagarwal@ualr.edu

T. Özyer et al. (eds.), *Social Networks and Surveillance for Society*, Lecture Notes in Social Networks, https://doi.org/10.1007/978-3-319-78256-0_2

1 Introduction

Technology is evolving and this evolution changed the way we access information, express our opinion, or communicate with each other. The inexpensive nature, easy to use, and popularity of social media made it a powerful tool that can be used to disseminate misinformation or coordinate cyber propaganda campaigns in order to achieve strategic and political goals, influence mass thinking, and steer behaviors or perspectives about an event. This motivated us to investigate these phenomenons and conduct this research/study. For example, there were a lot of anti-NATO propaganda that were pushed by pro-Russian media that presented NATO's exercise (Trident Juncture Exercise 2015) as preparation for WW3, or acts of provocation for Russia, and these exercises were not defensive in nature but openly aggressive. Such propaganda was shared in many social media outlets such as blog sites,[1] Twitter,[2] or YouTube channels.[3]

A study showed that cybercriminals tend to collaborate or even transact cyberattack tools via the "dark markets" that exist in the online social media [1]. In addition to cyber criminals, deviant groups can also collaborate and conduct acts collectively using this market. We define online deviant group (ODG) as a group of individuals that organizes a harmful activity using cyberspace in which its result would affect cyberspace, physical space, or both, i.e., the "cybernetic space" [2].

Social media such as blogs, Twitter, Facebook, Google Plus, Instagram, and the likes are rich sources of information [3]. With millions of social network users around the globe, cyber forensic analysis of social media has profound applications [4]. Cyber forensic analysis of social media sites can help collect evidence that helps investigators develop a strong case [3]. Cyber forensics is "the process of acquisition, authentication, analysis, and documentation of evidence extracted from and/or contained in a computer system, computer network, and digital media" [5]. With the use of metadata extracted using cyber forensics, such deviant groups can be discovered. We develop methodologies that can be used to identify such deviant groups.

Digital forensics research can be categorized into two themes, *theoretical* and *technical*. Theoretical research covers the development of theories and methodologies including models, frameworks, and processes to conduct digital forensics investigations. A research was done by Lau et al. [1] in which they applied a probabilistic generative model to mine cyber criminal networks from online social media. Their model outperformed the support vector machine (SVM)-based method (by 16.62% area under the ROC curve) and the latent Dirichlet allocation (LDA)-

[1]Trident Juncture: NATO's Largest Military Exercise since Cold War. The "Fictitious Target" is Russia (GlobalResearch.ca, available at http://bit.ly/294Uo2E)

[2]War game whoops! NATO exercises end with hovercraft, Humvees stuck in sand (VIDEO) (RT.com, available at http://bit.ly/298ya0R)

[3]YOU WON'T BELIEVE WHAT NATO IS DOING TO PREPARE FOR WORLD WAR 3! (YouTube.com, available at http://bit.ly/29bcH7P)

based method (by 5.23% area under the ROC curve) [1]. On the other hand, technical research covers the development of tools and techniques to help in digital forensics investigations [6]. Collecting social media evidence is different than traditional digital forensics, which usually requires investigators to extract data from a piece of hardware in the possession. Social media providers like Facebook or Twitter are not going to help investigators get that evidence unless it is an extreme case [3, 4]. One of the problems in cyber forensics is how to visualize the collected cyber forensics data in an easy-to-understand format [4]. Some techniques an investigator might use is screencast tools, such as Microsoft's Skydrive or Screencast-o-matic, which record whatever an investigator might see on social media [3].

In this work, we identify and study the behavior of coordinating deviant groups in social media during the cyber propaganda campaigns using social network analysis and cyber forensics techniques. For cyber forensics data, we use Maltego (available at: https://www.paterva.com/web6/products/maltego.php), which is a tool that can be used to gather any publicly available data that might provide an insight into how different social media platforms (e.g., blog sites connected to Twitter accounts) are connected or affiliated [7]. We use computational social network analysis in combination with the metadata extracted from the cyber forensics tool to have a comprehensive understanding of the entire propaganda campaign coordination.

For conducting social network analysis we use NodeXL [8] and focal structure analysis (available at http://www.merjek.com). Focal structure analysis was implemented by Sen et al. [9] to discover an influential group of individuals in a large network. These individuals are connected and may not be the most influential individually, but by acting together they form a compelling power. This approach was tested in many real-world cases including the Saudi Arabian women's right to drive campaign (Oct26Driving campaign) on Twitter, and the 2014 Ukraine crisis when President Viktor Yanukovych rejected a deal for greater integration with the European Union [9]. For blog analysis, we use Blogtrackers (available at http://blogtrackers.host.ualr.edu/) [10].

The implication of this research is interesting not only for the scientific community, but also for authorities as these deviant groups pose non-negligible concerns for public safety and national security; for example, in many cases these groups call for civil disobedience, planned protests, or direct actions against specific events. Therefore, in this study, we propose to seek answers to the following questions that further help us analyze cyber propaganda campaigns:

1. Are there any blogs used by the groups to disseminate propaganda? How can we identify those blogs?
2. Who are the most coordinating/influential groups in the network? Which nodes are the most communicative or most powerful to disseminate the message using their social ties? What are the most used platforms by the individuals in the group?
3. How is the propaganda resonating with the community? What is the public opinion mostly concerned about? What are the top tweets, top hashtags, etc.?

4. Who are the most important individuals in the network (active tweeters, most communicative nodes in the network)? Can we identify influential narratives in the cyber campaign?
5. Can we identify the coordinating bots used during the propaganda campaign and study their content and behavior?

Seeking answers to the aforementioned questions, we make the following contributions in this chapter:

- We design social network analysis and social cyber forensics informed methodologies to study the sociotechnical behaviors of individuals during cyber propaganda campaigns to develop detection tools ready to be deployed for cyber operations.
- We provide a deeper understanding of social media as a facilitator for a group's activities to advocate a specific agenda.
- We were able to identify influential users on Twitter and blogs and find the relationship between them to study the cross-influence of various social media platforms in conducting strategic information maneuvers during cyber propaganda campaigns.

Rest of the chapter is organized as follows. Theoretical background of the research is discussed in Sect. 2. Section 3 presents the research methodology including two datasets used to study the cyber propaganda campaigns with different methodologies followed with each. Analysis done by our developed Blogtrackers tool is presented in Sect. 4 to examine the methodologies' efficacy. Section 5 concludes the chapter with possible future research directions.

2 Literature Review

In this section, we review some of the literature that is relevant to this research and discusses how our work is different than the others. In Sect. 2.1, we discuss some of the literature that has been done in the cyber forensics field, particularly the evolution of the usage of digital forensics tools over time; then we discuss what the term "data carving" is; and finally, we discuss the forensics tools used to collect data from social networks. In Sect. 2.2, we review some of the previous work done related to the usage of bots in information operations. In Sect. 2.3, we review some of the work done on calculating the influence score in social media channels, e.g., Twitter and blogs.

Cyber Forensics

For the last three and a half decades digital forensics tools have been evolved from simple tools, which were used mainly by law enforcement agencies to important tools for detecting and solving corporate fraud [11]. Cyber forensics tools are not a new type of tools but they are evolving over time to have more capabilities, more exposure to the audience (investigators or public users), and the type and amount of data that can be obtained by using each tool. Cyber forensics tools can be traced back to the early 1980s when these tools were mainly used by government agencies, e.g., the Royal Canadian Mounted Police (RCMP) and the US Internal Revenue Service (IRS), and were written in `Assembly` language or `C` language with limited capabilities and less popularity. As time passed these tools got more sophisticated and in the mid of 1980s these tools were able to recognize file types as well as retrieve lost or deleted files, e.g., *XtreeGold* and *DiskEdit* by Norton. In 1990s these tools become more popular and also have more capabilities; for example, they can recover deleted files and fragments of deleted files such as *Expert Witness* and *Encase* [12]. Nowadays, many tools are available to the publics that enable them to collect cyber forensics data and visualize it in an easy-to-understand way, e.g., Maltego[4] tool.

Data carving is a term widely used in the field of cyber forensics which means "Identifying and recovering files based on analysis of file formats" [5]. Carving digital forensic data can play a vital role in solving digital crimes. A lot of research has been conducted to improve data carving [5]. According to Nadeem et al.'s [11] literature review on articles that are related to the digital investigation or computer forensics on *disk area*, more research is needed to improve data carving techniques to retrieve important data and evidence from damaged or corrupted data resources [11]. Although there is a big research thrust on *data carving* within the forensics area, this is not exactly relevant to our research efforts, as we are not collecting forensics data from a disk or a device instead collecting metadata using forensics tools (i.e., Maltego) and techniques (e.g., websites' unique identifiers, e-mail addresses, or IP addresses among others).

Social network forensics tools collect data in many different ways, e.g., crawling by using the social network APIs, extract artifacts from local web browser cache, sniffing on unencrypted Wi-Fis (active attacks) or with ARP spoofing on LANs, or using a third-party extension for the social network in combination with a traditional crawler component (friend in the middle attack) [4]. Another investigative tool that might be used is to hook into the APIs of social media such as Facebook or blogs and collect metadata of a blog entry or Facebook walls such as time stamps, affiliations, IP addresses, locations, or e-mail addresses [3]. Research by Noora et al. [13] obtains cyber forensics evidence from social media applications that are installed on smartphones. Their research was testing whether the activities conducted through

[4]Maltego, available at www.paterva.com/web6/products/maltego.php

these applications were stored on the device's internal memory or not. They used three major social media apps, i.e., Facebook, Twitter, and MySpace, and three device types, i.e., iPhone, BlackBerry, and Android, for their experiments. The results show that BlackBerry devices do not store any information that can be retrieved by digital forensics tools while iPhone and Android phones store a significant amount of valuable data that can be retrieved [13]. Some work focused on extracting forensics data of social media apps from the computer hard disk, e.g., carving artifacts left by the use of Facebook Chat on a computer's hard disk [14]. A novel method to harvest/collect/carve data from social network websites, e.g., Facebook, was introduced by Markus et al. [15]. They designed a hybrid system that is based on a custom add-on for social networks in combination with a web crawling component. Their system is able to carve "*social snapshots*" which are defined as all the profile information of the target user, e.g., user data, messages, photos, and the associated metadata, e.g., internal timestamps and unique identifiers [15]. In our research, we are not creating a tool to collect forensics data from social networks; instead we are using an open-source tool called Maltego which is developed by Paterva.[5] Maltego is an open-source tool that collects Open Source Intelligence (OSINT) and forensics data. This tool provides a library of transformations for the discovery of data from open sources. It helps analyze the real-world connections between groups, websites, and affiliations with online services such as Facebook, Flickr, and Twitter. It also provides the capability to extract and visualize the results in a graph format that is suitable for link analysis.

Empirical Observations and Trends of Bots Used in Information Operations

It has been widely observed that in a cyber propaganda campaign, alongside human actors, bots/botnets/automated social actors (ASA) also participate in the dissemination of propaganda. It is, therefore, important to study the types, categories, and strategies that bots use to act as powerful propaganda dissemination tool during crises. An earlier study [16] reported that bots were used in a sophisticated manner to disseminate propaganda during the uprising of the Syrian civil war. The authors studied the "Syrian Social Bots" or SSB examining the content of the tweets and categorized the bots based on the content posted, the time before they get suspended, and the type of activity the bot does (tweet or retweet). The study classified the bots into the following types:

- *Core Bots which* have three categories:
  1. *Short-Lived Bots*: *retweet* a lot but seldom *tweet* and they lasted for *less than 6 weeks* before Twitter suspended the account.

[5]Paterva (Pty) Ltd. a new train of thought, available at www.paterva.com

2. *Long-Lived Bots*: *retweet* a lot but seldom *tweet* and they lasted for *more than 25 weeks* before Twitter suspended the account.
3. *Generator Bots*: *tweet* a lot but seldom *retweet* anything.

- *Peripheral Bots:* accounts that participate in the dissemination process. Their task was *retweeting* one or more tweets *generated by the core bots.*

We did further investigation and studies to identify the types and strategies these bots use, and categorize them. Towards this direction, we studied the bots used to disseminate ISIL's beheading video-related propaganda (i.e., the Egyptian Copts, the Arab-Israeli "Spy," and the Ethiopian Christians) [17, 18]. We also studied the bots used to disseminate propaganda during the Crimean water crises [19] and the Dragoon Ride exercise [20]. We studied two different bipartite networks, i.e., User-Texts and User-URLs, in addition to one-mode networks, i.e., communication network (tweets, retweets, and mentions network) and social network (friends/followers network). These studies resulted in the following findings:

- Botnets were used heavily in all the cases to disseminate propaganda.
- The tweets contain mainly hashtags or keywords that may or may not be related to the event and URLs that point to resources related to the propaganda (e.g., videos, images, and memes).
- Accounts were identified as *bots* if they have the following characteristics:
  - The content of the tweets contains a high frequency of URLs.
  - The names of the accounts are very similar.
  - Tweets a lot in short period of time.
  - The tweets contain characters usually a human user would not post.

We were also able to identify some of the information maneuver strategies that botnets use to disseminate their propaganda such as the following:

- *Misdirection:* a technique used by magicians to make the crowd look somewhere else while they are performing the trick. For example, the bot would tweet unrelated news that is happening somewhere else but still mention a hashtag related to a crisis.
- *Smoke Screening*: when a bot would mention something about, for example, Russia or Ukraine, but not necessary related to the crises [19]. Similar techniques have been used in the Syrian Social Bot (SSB) to raise awareness of the Syrian civil war [16].
- *Thread-Jacking*: the change of topic in a "thread" of discussion in an open forum.
- *Hashtag-Latching*: strategically associating unrelated but popular or trending hashtags to target a broader, or in some cases a very specific, audience.

Influence in Social Media

Blogs provide a rich medium for deviant groups to build and frame propaganda by using half-truth or twisting facts to influence the masses. On the other hand, Twitter is limited by its number of characters (i.e., 140 characters) and is merely used as a dissemination medium or communication network. Instead, Twitter is used as a driving vehicle to steer the traffic of its audience, i.e., followers of the account to the blog sites where the bloggers have no limitation on words for framing and disseminating propaganda over social media. It is important to understand the disinformation dissemination network on Twitter, but it is more important to understand the activity generated by the blog posts or the bloggers.

Identifying influential individuals is a well-studied problem. Many studies have been conducted to identify the influence of a blogger in a community [21–25]. The basic idea of computing the influence a blogger has is to aggregate the influence of their individual blog posts. If a blog post has a lot of in-links and comments then it indicates that the blogosphere has interest in this blog post. In-links and comments contribute positively towards the influence of the posts whereas out-links of the blog posts contribute negatively towards the influence. Hence, one way to compute influence is by a weighted linear equation of in-links, comments, and out-links of a blog post [22]. An alternate approach is to use a modification of Google page rank to identify influential posts as well as bloggers [24].

3 Methodology

In this empirical study of Anti-NATO and Anti-TRJE 2015 cyber propaganda campaign, we studied two types of datasets collected from two different sources, i.e., a dataset we collected using Twitter API through NodeXL and a dataset collected by Scraawl.[6] For each of these datasets, we followed a path of analysis based on the information available. The two paths of analysis are connected to have a comprehensive understanding of the cyber propaganda campaigns that were projected against TRJE 2015 exercise and NATO. Figure 1 depicts a flowchart of the methodology we followed in this study.

[6]Scraawl, available at www.scraawl.com

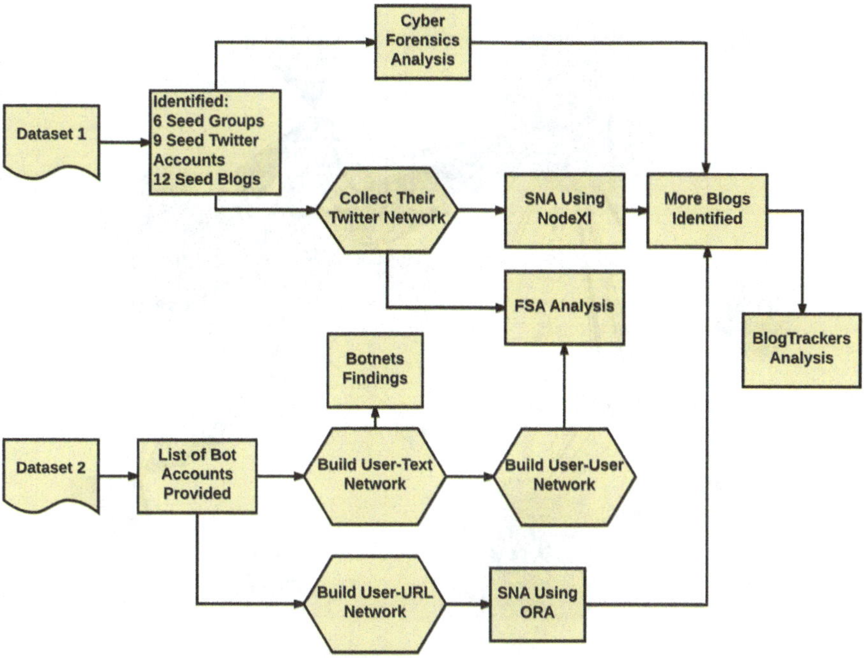

Fig. 1 Methodology to study the anti-NATO propaganda campaign

Dataset 1: Twitter Accounts of Historical Deviant Groups Known for Their Anti-NATO Narratives

Dataset Description

We identified six groups that propagate their messages on social media inviting people to act against NATO and TRJE 2015 exercise. An initial set of 12 blog sites were identified that the groups used to develop narratives against the TRJE 2015 exercise. We were also able to identify Twitter handles used to steer the audience from Twitter to their blogs. We identified an initial set of nine Twitter accounts used by the six groups. We used Twitter API through NodeXL to collect a network of *replies, tweets, mentions, friends,* and *followers* for all the nine Twitter accounts and whoever is connected to them with any one of the aforementioned relationships for the period 8/3/2014 4:51:47 PM UTC to 9/12/2015 3:22:24 AM UTC. The dataset file we obtained contains 10805 friends/followers, 68 replies, 654 tweets, 1365 mentions, 9129 total nodes, and 10,824 total edges (see Fig. 2).

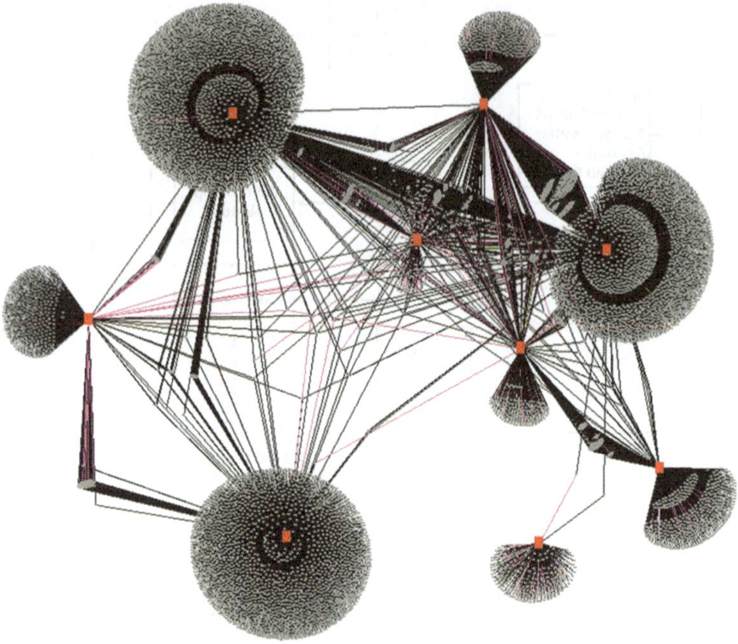

Fig. 2 Twitter social network of the nine users identified from cyber forensic analysis. The nine users identified are indicated in red color and all other users are indicated in gray color. *Friends and followers* edges are depicted in black color, *mentions* relation is depicted in purple, *replies* are depicted in green, and *tweets* are in gray

Metadata Extraction Using Cyber Forensics Approaches

By using the Unique Identifier, e.g., Google analytics ID obtained by Maltego, we can infer connections among blog sites. Maltego is an open-source intelligence and forensics application. It saves a lot of time in mining and gathering of information as well as the representation of this information in an easy-to-understand format. *Google Analytics* is an online analytics tool that allows a website owner to gather statistics about their website visitors such as their browser, operating system, and country among other metadata. Multiple sites can be managed under a single Google analytics account. The account has a unique identifying "UA" number, which is usually embedded in the website's HTML code [7]. Using this code other blog sites that are managed under the same UA number can be identified. This method was reported in 2011 by Wired, and also was cited in the book titled "Open Source Intelligence Techniques" by Michael Bazzell, an FBI cyber crime expert [7, 26].

We used a seed set of 12 blog sites to discover other blogs that are connected to them using Maltego as explained earlier. We used Maltego in a snowball manner to discover other blog sites. We were able to identify additional nine blogs that are connected to the initial seed blogs by the same Google analytics IDs. These

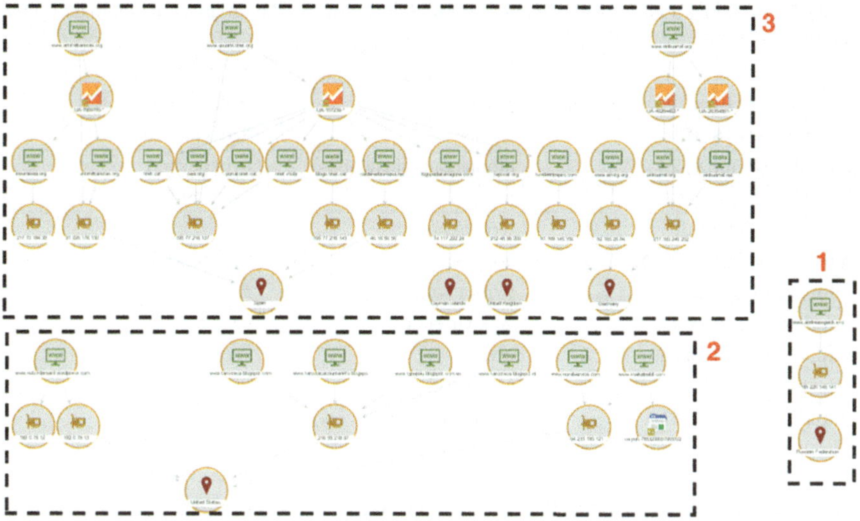

Fig. 3 Additional blog sites were identified using Maltego. Finding the IP addresses of all the blogs and their location gave us three clusters. Cluster 1 represents 1 blog site located in Russia, Cluster 2 represents 8 blog sites located in the USA, and Cluster 3 represents 16 blog sites located in Spain, Cayman Islands, the UK, and Germany

newly identified websites have the same content published on different portals and sometimes in different languages. For example, a website written in English may also have another identical version but written in another language that is native to the region. Such blogs are also known as *bridge blogs* [27]. We went a step further to collect the IP addresses, website owner name, e-mail address, phone numbers, and locations of all the websites. We obtained three clusters of websites based on their geolocation (see Fig. 3). These clusters are helpful to know the originality of the blog sites, which would help an analyst understand the propaganda that is being pushed by the specific blog site. Cluster 1 contains 1 website that is located in Russia, Cluster 2 has 8 websites located in the USA, and Cluster 3 has 12 blog sites located in Spain, Cayman Islands, the UK, and Germany. The result is shown in Fig. 3. From initial 12 blog sites, we obtained 21 blog sites, 6 locations, and 15 IP addresses. All the blog sites we identified during this study were crawled and their data was stored in a database that the Blogtrackers tool can access and analyze.

Applying Social Network Analysis to Identify Influential Information Actors

After finding other related blog sites used by the group to disseminate their propaganda using cyber forensics tool (Maltego) and methodology (unique identifier) we applied social network analysis to find who are the most important nodes in the

Table 1 The most used URLs, domains, and hashtags used by the group to disseminate propaganda

Top URLs in tweet in entire graph	Top domains in tweet in entire graph	Top hashtags in tweet in entire graph
http://bit.ly/1TSo024	fb.me	#cosymposium
http://bit.ly/1mpX3Z5	utopiacontagiosa.org	#yoconvoco
http://bit.ly/1Fhmxtw	bit.ly	#sinmordazas
http://bit.ly/1JeDzeL	wri-irg.org	#co
http://bit.ly/OFGM2015	com.es	#leyesmordaza
http://ow.ly/HcVkx	twitter.com	#siria
http://bit.ly/1dhJBkL	sinkuartel.org	#cos
http://j.mp/llamamiento	ow.ly	#coday
https://diasp.eu/posts/3493166	eldiario.es	#Egypt
http://bit.ly/1UFcIPD	diasp.eu	#Turkey

entire graph by activity type. We also wanted to know the most used hashtags during the time of the exercise. This can help in targeting the audience who follow that hashtag if counter-narratives were necessary to be pushed to the same audience; the most tweeted URLs in the graph which gives an idea about the public opinion and concerns; and the domains used most in the entire graph that helps to know where the focus of analysis should be directed, or what other media platforms are used (see Table 1). Using NodeXL we were able to answer all the aforementioned questions, for example, two of the top ten hashtags used during the TRJE 2015 exercise were #YoConvoco (using Google translation service to English as "I invite") and #SinMordazas (using Google translation service to English as "No Gags"). These two hashtags were referring to a campaign that is asking people for protests and civil resistance or civil disobedience. Also, investigating the top ten URLs shared the most in the dataset collected reveals that these URLs were links to websites that are asking people for fiscal objection to military spending on wars from the income tax return. Exploring the top domains also help us identify more blogs to crawl that disseminate propaganda against TRJE 2015.

Applying Focal Structure Analysis (FSA) to Identify Powerful Groups of Individuals Effecting Cyber Propaganda Campaign

To study the cyber propaganda campaign further, we apply focal structures analysis (FSA) approach to find the influential group of nodes. To find the most influential group of nodes in the network we divided our data file (9129 nodes and 10,824 unique edges) into two types of networks namely "social network," derived from friends and follower's relations and "communication network," derived from replies and mentions relations. We ran the FSA approach on these two networks to discover the most influential set of nodes or the seeders of information in the community.

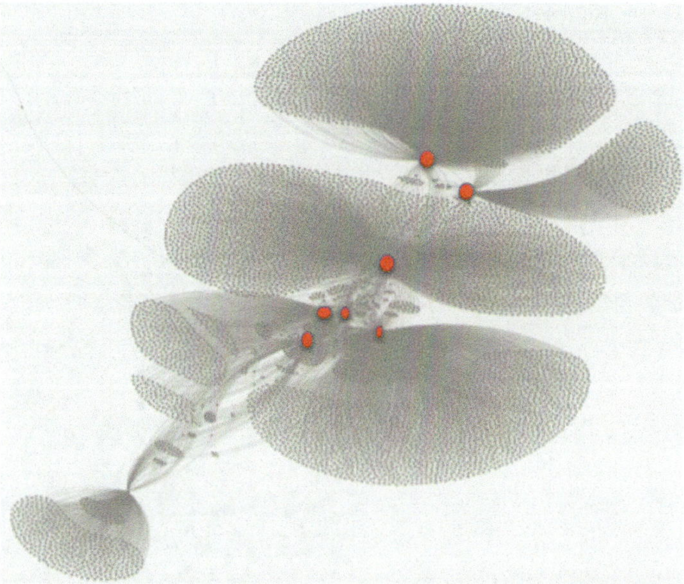

Fig. 4 Twitter social network of the nine users with members of the single most important coordination structure indicated in red color

Running FSA on the *social network* resulted in one focal structure with seven nodes (see Fig. 4). These seven nodes are in fact among the nine anti-NATO seed nodes we started with and are very tightly knit (i.e., they exert mutually reciprocative relationships). This indicates a strong coordination structure among these seven nodes, which is critical for conducting information campaigns.

Running FSA on the *communication network* resulted in 3 focal structures with a total of 22 nodes (see Fig. 5). The same seven accounts (out of the nine seed accounts) found in the social network focal structures are distributed in these three focal structures. This gives those seven accounts more power/influence than other nodes in the network because they are found in the focal structures of both networks, i.e., the communication and social networks. The rest of the nodes (i.e., the additional 15 accounts) found in these 3 focal structures of the communication network are new nodes. These are important because they are either leaders or part of key groups conducting propaganda campaigns.

Summary of the Conducted Analysis on Dataset 1

- We collected the Twitter network an "Agent X Agent" network (both the "communication network" and the "social network") of the six seed deviant groups who had nine twitter accounts. Then we analyzed this network to discover who are the agents/accounts/nodes that are top ranked in their activity, i.e., tweet,

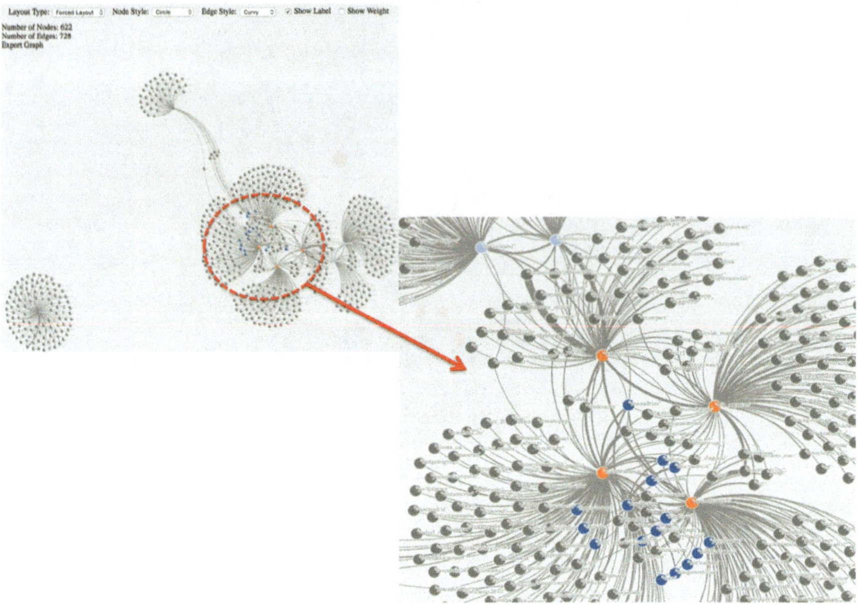

Fig. 5 Communication network (mentions and replies networks). Focal structures analysis approach helped in identifying a highly sophisticated coordinating structure, which is marked inside the red circle in the figure on top left. Upon zooming in on this structure (displayed on the bottom right), 3 focal structures were identified with 22 nodes. The color of the nodes represents the color of the focal structures

retweet, or mention the most. We also discovered the most used hashtags, the most tweeted URLs in the graph, and the domains that are used the most in the entire graph. This served as a *node-level analysis*.

- Then we used cyber forensics tool and techniques to discover the hidden relationships between the other blog sites that are related to the seed blogs.
- Then we applied focal structures analysis (FSA) to discover the coordinating groups. This served as *a group-level analysis*.

Dataset 2: Twitter Accounts of the Anti-NATO Bots Used During the Cyber Propaganda Campaign

Dataset Description

This dataset was collected using the Scraawl tool. It was noticed that bots were used to speed up and amplify the propaganda campaigns. Scraawl tool identified around 218 bot accounts. Botnets' tweets, mentions, and retweets were collected. The data was collected for the period from October 8, 2015, to October 11, 2015, that resulted in 869,062 tweets, 37,042 mentions, 74,898 retweets, and 308 unique users.

Applying Social Network Analysis to Identify Influential Bots

We constructed two types of networks, viz. User-URLs (two-mode network, i.e., “Agent by Knowledge” network where agent is the Twitter account and the knowledge, in this case, is the URLs contained in the tweets) and User-Texts (two-mode network, i.e., “Agent by Knowledge” network where agent is the Twitter account and the knowledge in this case is the whole text collected including the tweets, retweets, and mentions). For the ***User-URLs*** network, we used ORA NetScenes[7] to extract the top 20 tweeted URLs, mentioned URLs, and retweeted URLs. This network enabled us to find out that:

- Many of the URLs in the top 20 **Mentioned** URLs were mainly talking about NATO and the TRJE 2015 (propaganda dissemination). So more attention should be given to the *mentions network*.
- We identified two of the top 20 **Tweeted** URLs as interesting URLs because these websites show Ukraine and Russian Trending Topics/Hashtags on Twitter. This is an indication of tools that these bots use to get the most trending hashtags associated with the content of their tweet so they can reach wider audience (Hashtag-Latching).
- Most of the top 20 **Tweeted** URLs were NOT of relevance to NATO or the TRJE 2015. Most of the URLs were spam; e.g., gambling websites, coupons websites, food recipes, Android TV app, or a Turkish website helps users gain followers.
- The top 20 **Retweeted** URLs were NOT of relevance to NATO or the TRJE 2015. Most of these URLs were from Greek or Italian news websites talking about the Turkish election.

We used the blogs identified above as our initial seed of blogs to discover other blogs that are related/affiliated with them using Maltego. Then all these blogs are crawled and fed to Blogtrackers tool to perform more analysis. We also constructed a ***User-Texts*** network and we discovered the following:

- Many users are **tweeting** the same tweet content. This is considered as unusual behavior as it is almost impossible for two people to write the same exact text without a single punctuation difference unless they have a prior communication or same source of message. This behavior is an indication of “Clone Accounts” or “Echo Chamber.”
- We also have noticed that some of the bots that participated in TRJE2015 propaganda dissemination also participated in the past in other crises such as

[7]ORA NetScenes, available at http://bit.ly/27fuHnv

the Crimean Water Cries and the propaganda disseminated during the Dragoon Ride Exercise [20].

- Most of the strategies mentioned in Sect. 2.2 were used in disseminating the propaganda, e.g., Hashtag-Latching.
- Some users were repeatedly top ranked in the node-level measures calculated by ORA, which are the following:

  - *Capability*: Detects entities with high or low degree relative to other entities.
  - *Centrality-Out-Degree*: For any node, e.g., an individual, the out-links are the connections that the node of interest has to other nodes. For example, imagine an agent by knowledge network where the number of out-links an agent would have is the number of pieces of knowledge it is connected to. The scientific name of this measure is out-degree and it can be calculated on any network. Individuals or organizations that are high in out-degree for knowledge have more expertise or are associated with more types of knowledge than are others.
  - *Centrality-Row Degree*: Number of ties to others. Row sums of adjacency matrix.
  - *Cognitive Distinctiveness*: Measures the degree to which each pair of agents has complementary knowledge, expressed as the percent of total knowledge.
  - *Cognitive Expertise*: Measures the degree to which each pair of agents has complementary knowledge, expressed as a fraction of the knowledge of the first agent.
  - *Cognitive Resemblance*: Measures the degree to which each pair of agents has the exact same knowledge. The number of knowledge bit normalizes each value.
  - *Cognitive Similarity*: Measures the degree to which each pair of agents has overlapping knowledge.
  - *Correlation-Distinctiveness*: Measures the degree to which each pair of rows has complementary data, expressed as *the percent of total data*.
  - *Correlation-Expertise*: Measures the degree to which each pair of rows has complementary data, expressed as *a fraction of the data of the first row*.
  - *Correlation-Resemblance*: Measures the degree to which each pair of rows has the exact same bits. The number of columns normalizes each value.
  - *Correlation-Similarity*: Measures the degree to which each pair of rows has overlapping data.
  - *Exclusivity*: Detects entities that have ties that comparatively few other entities have. Individuals or organizations that are high in exclusivity for knowledge are those that have expertise or are connected to types of knowledge that few others have.
  - *Exclusivity-Complete*: Detects entities that have ties that no other entities have. Individuals or organizations that are high in complete exclusivity for knowledge are those that have expertise or are connected to types of knowledge that no one else has.
  - *Exclusivity-Knowledge*: Detects agents who have knowledge that few other agents have.

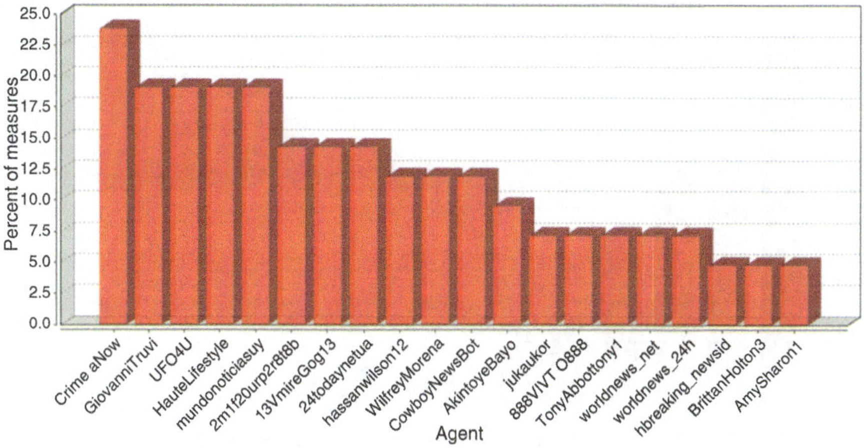

Fig. 6 The recurring top-ranked Agents chart shows the Agent that is repeatedly top ranked in the node-level measures mentioned above. The value shown is the percentage of measures for which the Agent node was ranked in the top three

This is an indication of how powerful these nodes are in the network. For example, we were able to identify many accounts that participated in other propaganda campaigns during other exercises, e.g., the Dragoon Ride Exercise. The ranks of these nodes are shown in Fig. 6.

Applying Focal Structure Analysis to Identify Powerful Bots Effecting Cyber Propaganda Campaign

We extracted the users mentioned and retweeted from the text to construct a one-mode (Agent by Agent) communication network (edges represent retweets, and mentions; note that the tweets were not included in this network because in this case they are self-loop edges and do not add any meaning in this case). This resulted in a network containing 18,987 nodes and 23,824 edges. Then we applied FSA algorithm to find the coordinating bots. This resulted in 15 focal structures (a strongly influential group of nodes) that include 286 nodes in total (see Fig. 7). Five accounts from the top 20 ranked nodes were also in the identified FSAs. This means that these accounts are not just top ranked in the Agent by Knowledge network but also coordinating accounts in the communication network (this gives them more importance).

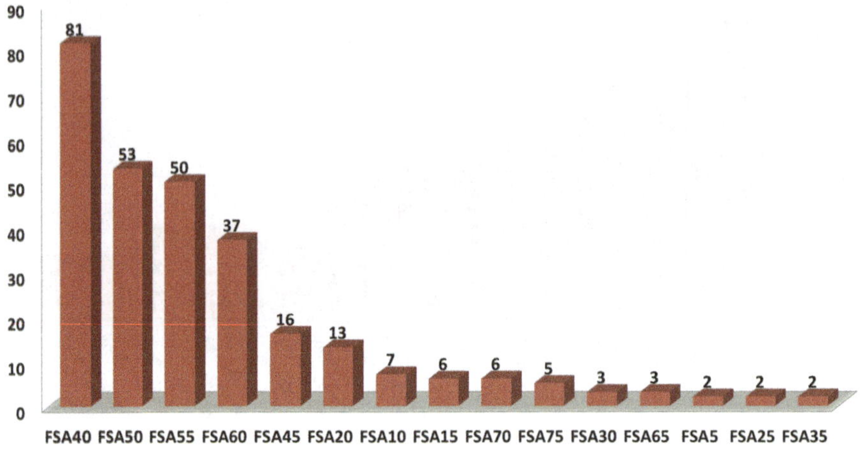

Fig. 7 The distribution of nodes among different focal structures. Here the FSA with ID = 40 has the highest number of nodes in it. The ID is used to distinguish each FSA and it is increasing in amount of 5

Summary of the Conducted Analysis for Dataset 2

- We created an "Agent by Knowledge" network to see who are the nodes/agents who are top ranked in the *node-level measures* aforementioned in section B. This gives us an insight into the important nodes in the network (or the active nodes during the propaganda dissemination). In other words, the nodes that share more knowledge (in this case tweet more), or nodes that mention more users in their tweets, have a higher participation rate in the discussion than other nodes. Also, the nodes that retweet a lot more than other nodes are those who are interested in spreading the messages to a wider audience. This serves as *node-level analysis*.
- We also created an "Agent by Agent" *communication network* and then we applied the focal structures analysis to discover the coordinating groups. It is important to know the important/influential/coordinating groups in the network. These nodes have so much power; by acting together they will be able to spread the message and be effective during propaganda campaigns. This is *a group-level analysis*.
- As expected some of the nodes that are top ranked in the "Agent by Knowledge" network were also in the focal structures. This gives these nodes more importance.

4 Blog Data Analysis and Findings

Dataset Description

Using the SNA and cyber forensic techniques mentioned in the previous sections we were able to identify more than 21 blog sites that disseminated propaganda against NATO and their TRJE 2015 exercise. We trained web crawlers using Web Content Extractor (available at http://bit.ly/1uUtpeS) to collect data from these blogs. Upon crawling them we obtained a total of 15,278 blog posts. These blogs were mainly located in the USA (we also have blogs located in Spain, Canada, Russia, Germany, the UK, and Palestine). The location of the blog site was determined either from the IP address location obtained from Maltego or if the author explicitly mentioned in his blogs where he/she is living. The majority of the blog posts were written in English language while posts in other languages were also collected, i.e., Spanish, Italian, German, Catalan, Arabic, French, Basque, Portuguese, and Russian.

Using Blogtrackers for Analysis

We started exploring the collected dataset by generating the *posting frequency* graph (depicted in Fig. 8) using Blogtrackers (available at http://blogtrackers.host.ualr.edu/), for the period of August 2014 to December 2015. We observed a comparatively higher activity in these blogs from September 2015 to December 2015, the period around the Trident Juncture Exercise (TRJE 2015).

We generated *a keyword trends* graph for the keywords "anti nato," "trident juncture," and "nato" using Blogtrackers (depicted in Fig. 9). The keyword trend for the "anti nato" completely aligned with the posting frequency graph in Fig. 8 indicating that the posts actually had "anti nato" keyword in it. We also observed that the trend for "anti nato" was consistently higher than "nato" for this time period indicating that there was more negative sentiment towards NATO in these blogs. We also found out that several blogs called for "anti-NATO" protests and movements.

We developed a *sentiment trend* (depicted in Fig. 10) for the same period that confirmed our findings; that is, more negative sentiment was observed than positive sentiment. Also, the number of blogs with negative sentiment was significantly more than blogs with positive sentiment.

We also ran the *Influential Posts* analysis [21, 22] in Blogtrackers to identify posts with high influence. Influence score of a post is computed using a stochastic model involving four factors, viz. *inbound links*, *outbound links*, *comments*, and *eloquence* of the blog post [21–25].

We found out that the *most influential posts* were from a blog site that was written in Italian language. Upon translation we found the post to be highly propaganda-riddled posts. It was demanding people to protest against the exercise conducted by NATO. The blogger of this blog used two of the conventional propaganda techniques

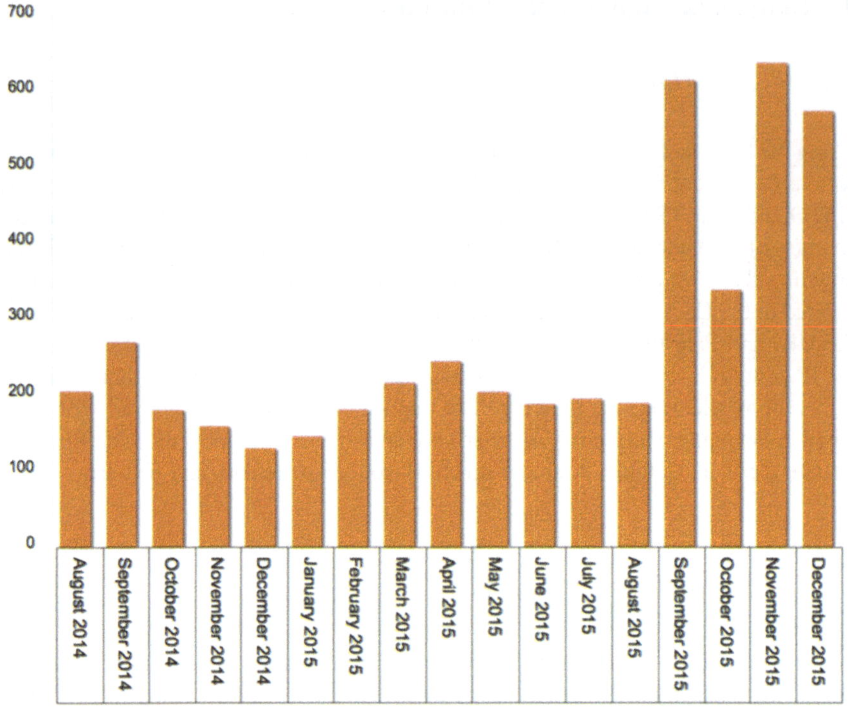

Fig. 8 The posting frequency graph generated by Blogtrackers shows an increase in the activity of blogs before TRJE 2015

called "Name Calling" [28] (associating a negative word to damage the reputation) and "Plain Folks" [28] (presenting themselves as ordinary people or general public to gather support for their cause or ideology). The blog post mentioned phrases like NATO exercise was contributing to pollution, and it was exploiting resources. It also categorizes this exercise as an act of militarization of territories to train for war.

5 Conclusion and Future Directions

In conclusion, the affordability and easy-to-use nature of social media gave it popularity to many people around the globe. The usage of social media meant to be for entertainment and to act as a mean of social communication environment but deviant groups harnessed this usage. Instead, they used it as a powerful tool to disseminate misinformation or coordinate cyber propaganda campaigns in order to achieve strategic and political goals, influence mass thinking, and steer behaviors or perspectives about some events. The latest behavior motivated us to do more

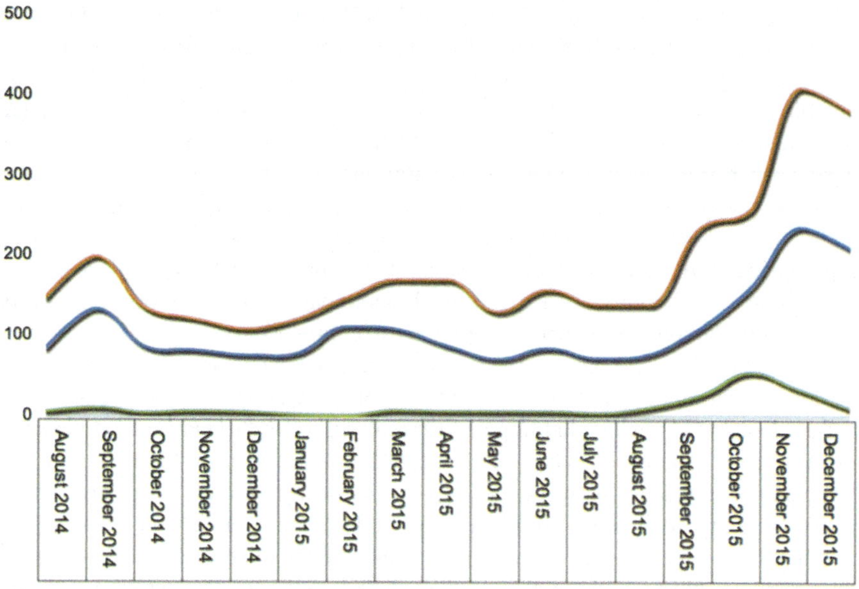

Fig. 9 Keyword trends for "anti nato," "nato," and "trident juncture" generated by Blogtrackers depicting the occurrence of these keywords over the time period

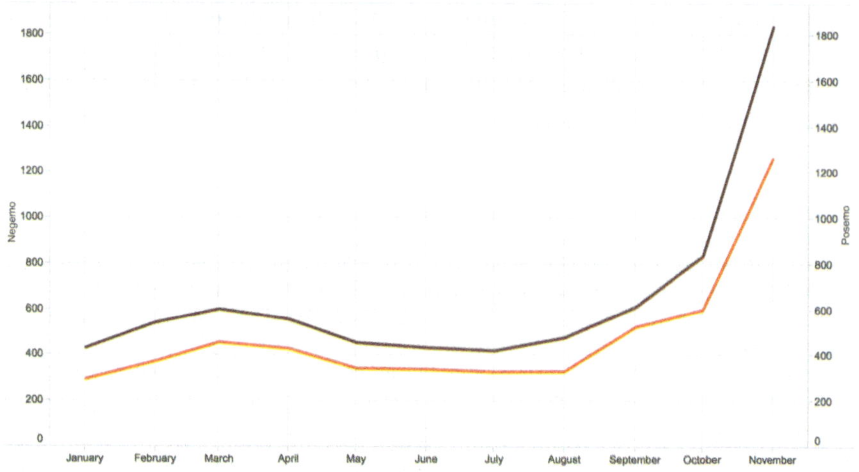

Fig. 10 Sentiment trends in the collected blogs

investigation, and study this phenomenon, and the behavior of the groups who conduct such acts.

In this work, we identify and study the behavior of coordinating deviant groups who created a lot of cyber propaganda against the TRJE 2015 on both Twitter and blogs. We utilized social network analysis, cyber forensics tools, and cyber forensics techniques to uncover the relation between the groups and to discover more groups.

We design these informed methodologies to help develop detection tools ready to be deployed for cyber operations. We were able to identify influential users on Twitter (who use Twitter as a tool to steer their followers on Twitter to their blog sites) and blogs and discover how they are connected/related to study the cross-influence of various social media platforms in conducting strategic information maneuvers during cyber propaganda campaigns. The aforementioned methodologies constitute a tiny but promising sample of an entire spectrum of approaches to extract metadata and relevant blog sites via cyber forensics. This set of methodologies would help in focused data collection, or guided snowball data collection, which then leads to the next phase of the study, i.e., a streamlined identification of key actors or bloggers that are top opinion leaders and top opinion disseminators.

For future work, we plan to focus on the messages that such groups are trying to spread. More precisely, we plan to do topic modeling of the messages to extract topical sentiments and targeted sentiments, to further enhance propaganda extraction. In addition to that, since some of the nodes that are top ranked in the "Agent by Knowledge" network were also in the focal structures, which gives these nodes more importance, we are planning to find out the nature of such nodes (i.e., are they bots or regular users?) and what makes them so special that they appear in both networks, e.g., what type of tweets they do (e.g., tweet news URLs, or use more hashtags than regular users), what type of activity (tweet, retweet, mention, or reply) they do most, and the frequency of tweets posted. Also, since blogs do not have any constraints on the number of characters unlike Twitter, they afford a platform conducive for framing narratives. We, therefore, plan to analyze blog-originated propaganda techniques from a communication and information science perspective for the modern social information and communication systems. Further, we plan to improve Blogtrackers' capability to scale with copious amounts of blog data and have more functionality.

Acknowledgements This research is funded in part by the U.S. National Science Foundation (IS-1636933, IIS-1110868 and ACI-1429160), U.S. Office of Naval Research (N000141010091, N000141410489, N0001415P1187, N000141612016, N000141612412, N00014-17-1-2605, and N00014-17-1-2675), U.S. Air Force Research Lab, U.S. Army Research Office (W911NF-16-1-0189), U.S. Defense Advanced Research Projects Agency (W31P4Q-17-C-0059), and the Jerry L. Maulden/Entergy Fund at the University of Arkansas at Little Rock. Any opinions, findings, and conclusions or recommendations expressed in this material are those of the authors and do not necessarily reflect the views of the funding organizations. The researchers gratefully acknowledge the support.

References

1. R.Y.K. Lau, Y. Xia, Y. Ye, A probabilistic generative model for mining cybercriminal networks from online social media. IEEE Comput. Intell. Mag. **9**(1), 31–43 (2014)
2. S. Al-khateeb, N. Agarwal, Analyzing flash mobs in cybernetic space and the imminent security threats a collective action based theoretical perspective on emerging sociotechnical behaviors. In *2015 AAAI Spring Symposium Series*, March 2015
3. B. Wright, Social media and the changing role of investigators. Forensic Mag. December, 2012

4. M. Mulazzani, M. Huber, E. Weippl, Social network forensics: tapping the data pool of social networks, presented at the 8th annual IFIP WG, 2012, vol. 11
5. D. Povar, V.K. Bhadran, Forensic data carving, in *Digital Forensics and Cyber Crime*, vol. 53 (Springer, Berlin, Heidelberg, 2011), pp. 137–148
6. S. Saleem, P. Popov, I. Bagilli, Extended abstract digital forensics model with preservation and protection as umbrella principles, presented at the 18th international conference on knowledge-based and intelligent information & engineering systems—KES2014, Gdynia, Poland, 2014, vol. 35, pp. 812–821
7. L. Alexander, Open-source information reveals Pro-Kremlin web campaign. (2015). Dostęp: https://globalvoices.org/2015/07/13/open-source-information-reveals-pro-kremlin-web-campaign/ Accessed 25 May 2017
8. M.A. Smith et al., Analyzing (social media) networks with NodeXL, in *Proceedings of the Fourth International Conference on Communities and Technologies,* 2009, pp. 255–264
9. F. Sen, R. Wigand, N. Agarwal, S. Yuce, R. Kasprzyk, Focal structures analysis: Identifying influential sets of individuals in a social network. Soc. Netw. Anal. Min. **6**, 1–22 (2016)
10. N. Agarwal, S. Kumar, H. Liu, M. Woodward, BlogTrackers: a tool for sociologists to track and analyze blogosphere. In *ICWSM*, March 2009
11. N. Alherbawi, Z. Shukur, R. Sulaiman, Systematic literature review on data carving in digital forensic, in *Procedia Technology*, 2013, vol. 11, pp. 86–92
12. K. Oyeusi, *Computer Forensics* (London Metropolitan University, London, 2009)
13. N. Al Mutawa, I. Baggili, A. Marrington, Forensic analysis of social networking applications on mobile devices. Digit. Invest. **9**, S24–S33 (2012)
14. N. Al Mutawa, I. Al Awadhi, I. Baggili, A. Marrington, Forensic artifacts of Facebook's instant messaging service, presented at the 2011 international conference for internet technology and secured transactions (ICITST), 2011, pp. 771–776
15. M. Huber, M. Mulazzani, M. Leithner, S. Schrittwieser, G. Wondracek, E. Weippl Social snapshots: digital forensics for online social networks, presented at the proceedings of the 27th annual computer security applications conference, ACM, 2011, pp. 113–122
16. N. Abokhodair, D. Yoo, D.W. McDonald, Dissecting a social botnet: growth, Content and Influence in Twitter, in *Proceedings of the 18th ACM Conference on Computer Supported Cooperative Work & Social Computing*, 2015, pp. 839–851
17. S. Al-khateeb, N. Agarwal, Examining Botnet behaviors for propaganda dissemination: a case study of ISIL's beheading videos-based propaganda, presented at the data mining workshop (ICDMW), 2015 IEEE international conference on, 2015, pp. 51–57
18. S. Al-khateeb, M. Hussain, N. Agarwal, Exploring ISIL cyber network activities: evolution, means, and strategies, in *XXXVI International Sunbelt Social Network Conference*, Newport Beach, CA, p. 4
19. S. Al-khateeb, N. Agarwal, Understanding strategic information Manoeuvres in network media to advance cyber operations: a case study analysing pro-Russian separatists' cyber information operations in Crimean water crisis. J. Balt. Secur. **2**(1), 6–17 (2016)
20. N. Agarwal, S. Al-khateeb, R. Galeano, R. Goolsby, Examining the use of botnets and their evolution in propaganda dissemination. Def. Strateg. Commun. **2**(2), 87–112 (2017)
21. N. Agarwal, H. Liu, L. Tang, S.Y. Philip, Modeling blogger influence in a community. Soc. Netw. Anal. Min. **2**(2), 139–162 (2012)
22. N. Agarwal, H. Liu, L. Tang, P.S. Yu, Identifying the influential bloggers in a community, in *Proceedings of the 2008 International Conference on Web Search and Data Mining*, 2008, pp. 207–218
23. S. Kumar, R. Zafarani, M.A. Abbasi, G. Barbier, H. Liu, Convergence of influential bloggers for topic discovery in the blogosphere, in *Advances in Social Computing*, 2010, pp. 406–412
24. A. Java, P. Kolari, T. Finin, T. Oates, "Modeling the spread of influence on the blogosphere," in *Proceedings of the 15th International World Wide Web Conference*, 2006, pp. 22–26
25. K.E. Gill, How can we measure the influence of the blogosphere, in *WWW 2004 Workshop on the Weblogging Ecosystem: Aggregation, Analysis and Dynamics*, 2004

26. M. Bazzell, *Open Source Intelligence Techniques: Resources for Searching and Analyzing Online Information*, 4th edn (CCI Publishing, 2014)
27. B. Etling, J. Kelly, R. Faris, J. Palfrey, Mapping the Arabic blogosphere: politics, culture, and dissent, in *Media Evolution on the Eve of the Arab Spring*, ed. by L. Hudson, A. Iskandar, M. Kirk (Palgrave Macmillan, New York, 2014), pp. 49–74
28. R.B. Standler, Propaganda and How to Recognize it, RBS0, 02-Sep-2005

Hiding Self on Facebook Sidebar Ticker, Surveillance, and Privacy

Chih Ping Chang

Abstract We could think of sidebar ticker as the same function of sidewalk, where people post messages, share information, or upload photos is just like they talk, photograph, or jog on the street. People don't feel being seen is offended when they are active down the street, but on social networks they feel different, mostly because they would expect Facebook as a virtual personal space. Unfortunately, it is public based on opinions of the court. Another concept to clarify is that public is not equivalent to "open to the public." Facebook might create a public space, and users should realize that the surveillance to them is the same fact on social networks as in society. However, users still have the right to decide their own conducts and exclude others from arbitrary interference, access, collection, and data usage.

The structure of this chapter is divided into four parts: ticker and privacy settings, surveillance on Facebook, discussion on privacy and public, and open to the public. In conclusion, we might expect to monitor others, hiding self behind the screen. We might feel safety hiding self in the crowd. However, social networks should be understood as a digitalized society with interaction, instead of a private space alone. Thus, we have to understand what real surveillance on social networks is. In this way, we can truly and effectively exercise the right to privacy.

Facebook began to provide a new service as showing live feed in the sidebar ticker in September 2011. Through this service, users can easily know their friends' activities and states on Facebook, such as "like it," "tag in," "check in," who they friend with, sharing, commenting and posting as well as its contents, and even those activities among friends' friends or groups, if they are open to the public. Some arguments arise; this phenomenon will be useful for cyberstalkers; this service provides an easy function to know interactions on Facebook not only among people we know, but also among people we don't know or not yet to know. Commenters are also worried that this will cause an unprecedented infringement of privacy, and even claim that

C. P. Chang (✉)
Shih Hsin University School of Law, Taipei, Taiwan
e-mail: ccp@mail.shu.edu.tw

T. Özyer et al. (eds.), *Social Networks and Surveillance for Society*, Lecture Notes in Social Networks, https://doi.org/10.1007/978-3-319-78256-0_3

the privacy is dead, because of strong connection of information by Internet and self-disclosure on social networking sites. They say that Facebook is like a huge panopticon [1, 2], where not only users are continuously watching and watched among them, but there might also be more objects, such as providers, aggregators, and governments, paying attention to our every move on Facebook. This is almost a transparent society [3]. Moreover, social networking sites, based on its powerful disclosure of information and social ties, made researchers recalculate everyone's degrees of separation number from the traditional 6 to 3.57 [4]. Even though we close our Facebook accounts, our information may be rebuilt from information released publicly by our friends' or family members' sites.

We may fear our privacy will be infringed through social network sites. One possible reason to explain such a fear is that, while we do online activities, we believe this Internet world is a virtual world, as an accepted assumption, not related to the real world. However, I have to point out that it ignores that the essential of the social networking sites is basically a digital model of our society. Taking this sidebar ticker as example, we can take it as the sidewalk by analogy. We can find people act on streets every day, where they are talking, photographing, eating, or jogging. No one think it is the invention of privacy that these activities are simply seen by others. On the contrary, those digital activities on Facebook sidebar ticker might be thought in a different way. People expect Facebook to be a virtual private space, and their digital activities or information should be considered as privacy. However, according to judgments by American and German courts,[1] since a social networking site exists for the purpose of social network sharing personal lives, it will be considered as a public space, unless there are specific privacy conditions [5].

Whereas the above statement is accepted, it does not mean that people can arbitrarily access to our information or digital activities uploaded on Facebook. I would like to clarify an important concept that must be clarified: personal information in public area does not equate that person wants to open information to the public. It is misunderstanding and inaccurate, for the ambiguity of privacy, that users choose to release their information on websites, and the web providers can collect, analyze, and reuse these data. General consent in terms of privacy is also a product under such myth. When you are jogging or eating ice cream on the street, it does not absolutely authorize someone else to take your pictures, record your actions, and then publish or edit them into films (or possibly add extra notes). The reason is the same. Facebook may create a public realm and users have to realize

[1]In the case of Katz v. United States, US Supreme Court held that once the information is exposed to the public, the person possibly no longer has a reasonable expectation of privacy. In the case of Romano v. Steelcase, Inc., the court indicated clearly that, because the most important function of SNS is sharing personal lives on the Internet, there was no reasonable expectation of privacy. German Federal Constitutional Court (BVerfG, Urteil vom 27.2.2008) recognized that there is no fundamental right to be interfered involved in this issue, when the police obtained the information from public Internet and open communications. The court also recognized another exception, in which the police directly participate in public communication networks and obtain something as evidences from there. No fundamental right to be interfered either.

that they are in fact under the social monitoring without any difference between real societies. Moreover, they still keep their rights against arbitrary interference and access by others in this public realm.

Therefore, this chapter discusses whether this Facebook ticker causes the invasion of privacy, based on the argument of panopticon or transparency society. In order to answer this question, this chapter first discusses Facebook ticker and its privacy settings, in order to clarify whether this fear of invasion of privacy is real. Then this chapter goes further to discuss a phenomenon of surveillance on Facebook and how to protect our privacy. Finally this chapter emphasizes this point of view once again, which is, personal information in public area does not equate that person wants to open information to the public.

1 Facebook Ticker and Its Privacy Settings

What Is This Sidebar Ticker

Ticker, as the sidebar, is located on the right of user's home page, to show users the things users can already see on Facebook, but in real time [6]. The user can see the whole news feed and join the ongoing conversation while he moves the mouse on the ticker. When there is a news feed, the ticker will be updated automatically, so that the user can better grasp activities of friends in real time. The user cannot close this service, but can hide the ticker or move it to the sidebar with chat room on the right.

Types of activities displayed by the ticker include the latest news stories, friends' status updates, likes and comments, and new photos from friends. Users in fact can find these activities and hints in other places on Facebook, such as home page or timeline. The function of the ticker is to reminder users what is going on among people whom they may concern.

"People can only see the posts, comments and likes that [the user] shares with them. So if [the user] share something with friends, only they will see it in ticker" [6]. The user can control what will be displayed on the ticker by "News Feed Preferences" function. If the user doesn't want to follow someone's posts or receive news from some Apps, he/she can hide these information. Hiding a friend's information or unfollowing his/her post just means his/her news won't be displayed on the user's home page instead of blocking this friend. The user still can interact with this friend. In another situation, the user can prioritize who to see first. News from these preferred friends will be marked and then Facebook displays them in particular for the user to see. Or the user can adjust numbers of friends to decide the amount of information displayed on his ticker. If the user sets the large number, he/she will receive news feed from almost all of his/her friends; but if he/she puts the very small number, he/she will only receive news feed from those most interested friends decided by Facebook.

This sidebar ticker is just used to integrate and filter information which was originally open to the public. In fact, the ticker is the Bata version of Facebook home page message, which combines news feed, friends' status, and contacts, and transfers information more immediately. It is the way to optimize information: On the one hand, it provides the user what he/she most wants to see, and on the other hand it also provides the user what is happening right now.

Ticker and Users' Privacy Settings

The ticker just displays information which the user has been authorized to know and he/she also can access in other places on Facebook, instead of changing the user's privacy settings. Facebook claims information or every users' activities shown on the ticker already exist on Facebook [6]. Ticker won't create new information or collect unauthorized information. Information on the ticker is the same as the original posted. In theory, what is displayed on the ticker depends on the user's privacy settings. That is, if the user cannot access someone's posts because of the privacy settings (e.g., the user is blocked by someone), these posts won't be displayed to the user in anywhere on Facebook (e.g., the user's timeline, news feed, or ticker). But we cannot deny that this alert function on the ticker will make users notice some neglected messages in the past. Facebook responds to this situation, if the user may begin to find some messages never noticed on his ticker; the probable reason is the ticker functions to search messages easier [6].

The ticker won't change the user's privacy setting, and what the user can see on his/her ticker depends on the privacy setting. The user also can control who can see his/her activities or messages through setting his/her Facebook privacy. This privacy setting not just affects information shown on the ticker, but also applies this personalized privacy policy to regulate other Facebook services, such as timeline, profile, and news feed. Only people who can access the original information (e.g., posts, photos, messages, and other activities) can continuingly follow the ongoing reactions (e.g., comments, shares, and like/thumbs-up). For example, if the user response of a photo is limited to sharing among the family, the user's friends won't see this response on the ticker because these friends are not authorized to access the original photo.

Summarizing Facebook's design to this service, we can say that there is nothing new shown on the ticker, and that its content basically depends on the privacy settings. Therefore, if we don't want others follow our activities, such as who we friend with or update of self-relationship, we can change the privacy settings and personalize it for our best privacy and security. This sidebar ticker let us notice our information is overflowing and spread freely for a long time, and the only solution that Facebook provides is the privacy settings. So it is time to know the Facebook privacy settings.

Privacy Settings on Facebook

First, we should rightly understand social relations and concepts of visibility of message on Facebook. When a Facebook user sets his/her information in public or used the privacy setting by default, B, who is not the user's friend, also can see what the user posts on his/her own Facebook page (Fig. 1, Situation I). If the user is limited to share his/her information in the group of friend, B in theory cannot access the user's posts or photos (Fig. 1, Situation II). In situation I, the stranger B can access the user's information because the user adopts the default privacy settings, which is public. Then everyone can see the user's posts or photos. But in situation II, the user sets his/her privacy only share among friends or specific friends, and then this stranger B cannot absolutely access the user's information (not only posts or photos, but also comments or like/thumbs-up). This information flow will be cut off by A (friend). Such an operation restates what the social networking site claimed in its privacy policy; users can control personal information and have choices to decide their privacy settings [7].

Second, we will discuss the privacy settings on Facebook. This social networking site gives users the ability to decide who can "see" the user's messages, comments, photos, or other activities. In the section of privacy settings and tools, Facebook allows users to create groups in friends list and differentiate them to authorize different levels of the visibility to them. Basically, Facebook provides four options to users to authorize the visibility. These four options are public, friends, "only me," and custom. Every user can access the post, comments, sharing, and other activities, when the user sets his/her privacy in the level of public; on the contrary, only the user can access his/her own activities if the setting is on "only me." The user can limit information flow among friends, just like situation II. He/she can also adjust this friends group through custom setting, such as sharing with friends except acquaintance or with family. But if the user allows friends' friends to access his/her information, he/she will afford the risk of losing his/her control on information flow. Someone may make friend with your friends just for obtaining information in secret. For example, A and B were intimate, when A had a relationship with C. After some misunderstanding, A broke up with C and blocked him on Facebook. However, A found B and C are in relationship because B set everyone or friends can follow B's

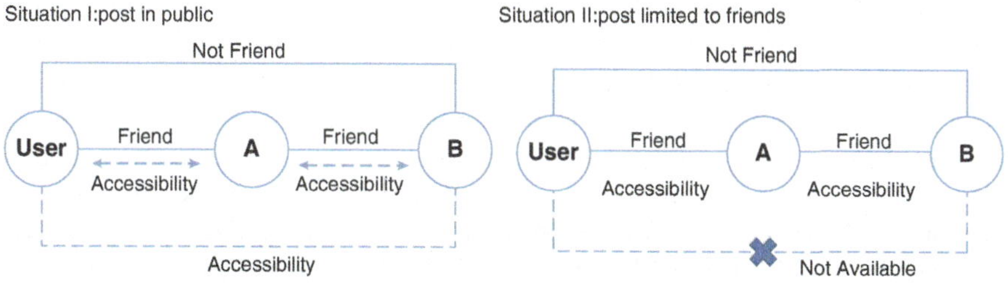

Fig. 1 Illustration of FB Privacy Settings

news. It seems Facebook disclosed B's betray to A. Users have to pay attention on their own privacy settings.

According to Facebook privacy settings, an assumption is behind, that is, every user is reasonable, acknowledges his/her privacy risks, and has the ability to control his/her information flow. Therefore, what Facebook provides is the protection of releasing information flow. In other words, the user is given the right to decide whether he/she wants to provide his/her information to others. However, if someone collects information with his/her consent and use it in other way (e.g., as criminal evidence), even though the user will be hurt or feel uncomfortable because of the use, this part of privacy is not protected under Facebook privacy settings.

To sum up, if we make good use of Facebook privacy settings, we can decide not being searched on Facebook; not releasing friends list; limiting information among intimate; not allowing others to see our posts or comments; not allowing friends to tag us in the photos, posts, or comments; and even not allowing friends post messages on our home page. The best way to protect our privacy on Facebook is individualizing our privacy settings instead of by default, if we don't want strangers to know our information.

2 Facebook Ticker and Surveillance

As many science and technology bloggers claimed, the function of ticker is to recommend users to filter information. Users intentionally upload information and ticker just integrates original information disseminated on Facebook. If ticker just functions as the filter using original information and Facebook privacy settings can protect users' privacy, why many people still concern their privacy will be infringed through the ticker? I boldly speculate that the reason is this integration function of the ticker makes information and digital activities visualized. First, activities in the Internet are known for anonymity. It is difficult to ask users to follow "real name" policy in social networking sites; even Facebook has failed [8]. People expect to hide themselves behind the screen. Second, peeping is human nature, which gives people a sense of control and pleasure of knowing everything [9]. Also hiding self on Facebook makes the user feel safe; he/she can peep, or speak bitterly in a virtual community and at the meantime log out from that community when he/she provokes every one. It is the resource for fun without the responding responsibility. Third, physical private space makes the user an illusion of privacy when he/she posts a message on Facebook in the room alone. In this situation people forget that there is a lot of information flowing on the Internet every second and it means that there are a lot of individuals accessing the Internet world from everywhere any time. Fourth and finally, people panic after realizing that they lost their control on information. People usually don't read Facebook privacy policy carefully. They just give their general consents for playing "Happy Farm." But the ticker shows their activities on the sidebar and they can do nothing. Fifth, the ticker displays people's activities as they are walking on the street. We can easily watch people's activities on the street. Even though we are playing Facebook or "phubbing"[2] alone in the room, our activities expose under eyes of the public and the ticker lets us to know this fact more clearly. We can watch we are being watched.

Therefore, people began to worry about the intervention of their privacy and further became concerned that their identities will be shaped by Facebook information feed, when Facebook provided this ticker service. A point of view was raised, "Facebook are the virtual Panopticons of our time" [2].

The Panopticon and Foucault's Interpretation

Panopticon, literally meaning a place where everything can be observed, now is to say a model prison designed by the utilitarian philosopher Jeremy Bentham. He designed a circular prison with cells arranged around a central well, from which prisoners could at all times be observed. On the contrary, prisoners could not watch the situations in the central well because of exposure to the strong light; thus they have no idea whether they are being watched, have been watched, or will be watched. Regardless of whether prisoners are actually being watched, this asymmetrical monitoring has led them to assume that they are being watched whenever they are.

Foucault uses the panopticon as a metaphor, although in history the panopticon has never been built; that is, it represents an abstract, obscure ideal power relationship rather than the power got through violence. The panopticon itself is not just a power mechanism, but also a constraint force, which is not only in prison, but also exists generally in various institutions of this modern society, such as hospitals, schools, and shopping malls. Foucault observed that inhabitants in the city, the same as prisoners in the panopticon, are well aware of having themselves been watched by others at any time. Moreover, this panoramic surveillance will leave its imprint on people's minds, have them to learn self-censorship, and thus constrain their own behavior. He called this effect as discipline. Although Foucault didn't mention surveillance in an information society when referring to the concept of discipline in the context of panoramic surveillance, he didn't deny that formation and accumulation of new technologies will expand the scope and effectiveness of monitoring. In history, techniques of discipline are created through the continuous evolution and the invention [10].

[2]"Phubbing is a term coined as part of a campaign by Macquarie Dictionary to describe the habit of snubbing someone in favor of a mobile phone." Wikipedia, https://en.wikipedia.org/wiki/Phubbing

Facebook as the Field of Monitoring

When the user wants to know what his/her friends are doing, he/she can easily open his/her Facebook home page, find their name in the friends list, and click their accounts, and then all their information on Facebook will display in front of him/her just in seconds. He/she can easily know who makes a new friend, who is traveling abroad, who just has a baby, and even who is now online the same as him/her. Through functions of check-in and tag-in, he/she can also follow that friends' whereabouts, even his/her friends are abroad. Facebook creates a similar situation with the panopticon: the user can watch people's activities on Facebook behind the screen through computers, smartphones, and tablets. People who act on the Internet won't know and has no way to know whether they are being watched. Facebook creates a simple environment for everyone to observe others. When ticker service is online, it makes this observation easier by integrating all information on the sidebar. Furthermore, users can highlight some friends as their preferences through privacy settings [6, 11], and the ticker will report every news of them at the first timing. Besides, users cannot stop this service. What they can do is to hide the ticker from their home pages, but as mentioned above, they are still being watched or will be watched.

3 Facebook Ticker and Privacy

In the context, this Facebook ticker indeed creates the risk that the user is exposed to being monitored. Then, this chapter discusses whether such a state will be considered as the invasion of privacy.

Privacy in Traditional Legal Approach

A classical definition of privacy is "the right to be let alone" advocated by the American judge Cooly [8, 9]. It means people need a separate space from others in order to develop self-identity and sound personality. As time goes, scholars are developing the concept of privacy, giving it more and more abstract meaning. For instance, a scholar defines privacy as the right "to determine for themselves when, how, and to what extent information about them is communicated to others."[3] Or a scholar claims privacy as general personality right (allgemeines Persönlichkeitsrecht) to protect private area and further develops the sphere theory,

[3]Reference to Wang, Ming-Li: Information Privacy: A Contextualized Analysis. (in Chinese) Chung Yuan Financial & Economic Law Review. 32, 59–105 (2014), at 63. But this sentence is from Westin, A. F.: Privacy and Freedom 7 (1967).

which divides sphere of personal activity into core sphere, intimate sphere, and social sphere, giving them protection in different levels [9]. Recently a scholar argues that privacy as information privacy is not only the right whether subjects determine their information open to others, but also the right that subjects can build their own personalities essentially [10]. There are many related researches on privacy, but the recognition of privacy is still very vague.

In practice, the "reasonable expectation of privacy" doctrine, built by Justice Harland in the concurrence in the American Supreme Court case of Katz,[4] is accepted commonly. This so-called reasonable expectation of privacy asks for a twofold requirement: the first is "a person has exhibited an actual (subjective) expectation of privacy" and, second, "that the expectation be one that society is prepared to recognize as 'reasonable'."[5] This two-prong doctrine intentionally divided public and private spheres (but not limited in the physical conception). We agree that there is privacy in a bedroom (in physical private area), but can we agree that privacy exists in public area? It is meaningful in this Katz case. Justice Harland defined the intervention of privacy from the view of subjects, changing his perspective from simply differentiate public/private realm. After all, privacy is for people building their personalities. Furthermore, the fact in the case of Katz is that the police monitor the suspect's conversation in public telephone booths. It involves whether there is privacy in a physical public area. The Katz may provide the more acceptable public/private dichotomy for protection privacy. In the meanwhile, the boundary line between public and private has become a more difficult problem to solve.

Privacy in Context

After Web 2.0 age, it becomes more and more complex to delineate this public/private boundary and difficult to find a stable standard. This is because information becomes the primary object of privacy protection, and its flow is difficult to define public and private. As discussed above, even though someone is using Facebook in his/her room, he/she may feel or be in the (virtual) public field, when he/she connects the social networking site. Thus, scholar coined three levels of privacy in information: public communications, private communications, and quasi-private communications. Communications in public area will not be protected, because they are open to share. If a person voluntarily provided information to others, there is no expectation of privacy in this case. On the other hand, communications in a private sphere means that information is only accessed by a limited group of people, such as e-mail, phone calls, chatting, or other instant messaging services. They are not intentionally divulged to the public, and it is likely

[4]Katz v. United States, 389 U.S. 347, 360 (1967).

[5]Katz, 389 U.S. at 361 (Harland, J., concurring).

to consider the protection of privacy in these communications. The level of privacy between the public and private communications is quasi-private communications, which are most difficult to classify. Take communications on Facebook as an example; user chooses to share information with friends, friends' friends, or other user-created networks, and communications cannot be accessible to the general community of Facebook, but are accessible to the user's friends or networks, which are called quasi-private communications. Although the information is not available to the general public, recent judges and legislators still think quasi-private communications as public, because a potentially large number of people will be able to access this information. Users may believe, on the contrary, that they have a reasonable expectation of privacy based on the privacy setting not allowing the public to access their Facebook [12].

However, Professor Helen Nissenbaum thought this boundary problem under the different approach. In her theory [15], protection of personal information should be considered with personal information flow in context at the time. When information transmits from one "information subject" to another, there should be different norms for information in order to correspond to cases in the different context. A context is broadly in our daily life, including situations among educational, political, economic, and social dimensions, and also including experiences on a variety of social structures in everyone's daily life. It is influenced by culture and history, natural environment, political structure, economic systems, religious institutions, and so on in a specific time and space in a specific society. Any context will follow the corresponding norms, and these norms will decide for individuals what the role to play, and what actions to take or be not taken. Norms for information are developed by relevant agents, type of information, and transmission principles. This theory jumps out from public/private dichotomy. In discussing the privacy issue on personal information, it argued that full account of the context of the situation should be taken, and observed as a whole [13]. While making decisions in our daily life, people often don't consider or depend upon only one factor, but all factors as a whole. A good example to explain this idea specifically is the case of the Burning Man Festival introduced by Professor Solov [14]. Participants can tolerate or even be willing to be taken their pictures by other participants, and to be shared/to share these photos in their group on the social networking sites. But these participants will feel that their privacy has been violated, while these photos were reprinted in newspapers or magazines or their websites. That is to say, privacy is floating, which will change due to information subjects, owners, and recipients under different backgrounds.

Under this approach of privacy as contextual integrity, scholar furthermore rethinks privacy towards its spatial connotations [16]. Connecting the classic conception of privacy, privacy should be a room, which is able to isolate our own thoughts from outside interference. Private space is the basic for building personality and completing our own thoughts. If a person cannot have the private sphere, his/her life will be completely exposed to others or under surveillance. This is exactly the panopticon. As Foucault's description, people become objects being watched inside the society, and their behavior were disciplined from outside the authority.

Finally they will lose their subjectivity. This room is not static, but changing with interactions among people and social concepts.

In my opinion, focusing on spatial connotations of privacy is workable. Privacy is actually the boundary dividing ourselves from others. This boundary may appear between physical sidewalk and bedroom, or between participants and journalists. The boundary is a line and the question is how to draw this line. Some people put this line more close to themselves, owing to the limited private sphere. They may not mind friends into their room and sit on their bed or look at his/her diary. Others put this line more strictly. They feel offensive while strangers are watching on their activities on Facebook. However, even as the line for a room, this boundary is blur, floating, and changeable in different cases. At present, the most appropriate solution to define this boundary is considering privacy in context case by case.

Does the Ticker Infringe Users' Privacy?

If we apply the "reasonable expectation of privacy" doctrine in the case of Facebook ticker, the second objective prong may not be satisfied. What is displayed on ticker comes from the original information posted on other Facebook services selected by algorithm. Such information can be found and has been disclosed by other services instead of the ticker. Showing users' state or activities on ticker doesn't create a new invention and break the general expectation. If we follow the trichotomy theory thinking of quasi-private communications, we may conclude that there is no invention of privacy. Information shown on the ticker is based on the user's privacy settings, which should be theoretically under his/her control. As Facebook claims, information will not be released to the unauthorized party. Therefore, it seems lack of necessity to protect privacy in this part. However, there will be prima facie evidence of intervention of privacy, if we examine this case under Nissenbaum's theory [13]. Facebook didn't obtain individual consent before putting this service online (and it is impossible in reality). Even though this service just integrates information distributed on Facebook, its function of notification may cause some known risks, such as relationship problem mentioned above or cyberstalking. And the user cannot refuse or close this service. Facebook only allows the user hide this service, which means if the user hides the ticker, he/she will not see the ticker on his/her screen, but other users still can follow his/her state or activities on their ticker. Although Facebook claims users can control their information through privacy settings and tools, for those who don't want to be followed in secret, the only choice in privacy settings is "only me." If someone has to do so, it is better he/she quit from Facebook.

I want to discuss this question in approach of the spatial connotation of privacy. If we think Facebook as a virtual society as real world, where people do their business and build a variety of relationships with others, the user's account will be like his/her room, his/her home page will be like the porch to connect outside, and this sidebar ticker will be the sidewalk in the real world. Most of us don't feel uncomfortable to be watched on the sidewalk, and it should be the same on the ticker. Of course, we have to consider the effect of notification, which will attract people's attention on information they ignored in the past. In fact, ticker displays excessive information and overloads people's minds. Most users will subconsciously ignore these huge amounts of messages to reduce the mental loading, just as what we do walking on the sidewalk. However, different with the sidewalk, this ticker not only displays users' state and activities, but also lets others know "who do this" by connecting users' accounts. What we see on the ticker is not someone is discussing this article, someone is talking to others, or someone is playing a game, but A comments on this article, B is talking with others (we may know the third party), or C is playing a game and gets some achievements. Walls for dividing outsides are broken by this visualization, when ticker names every state and activity, and makes information easy to follow. This is the reason that people feel fear when they find that it is easy to recognize others and dig what they are doing. They don't just feel or know they might be under watch, but they watch they are watched. We are also concerned with other kinds of notification as the same. Ticker makes a hole to peep others, connecting a virtual activity to a real person. We realize clearly that it is one of our friends doing this by watching on the screen. Unfortunately, we cannot reduce this risk through Facebook privacy settings and tools. With the plain words, Facebook only provides users whether they agree with Facebook to dig a hole on their wall and how big the hole is (only for friends or everyone can pass). Thus, users have to realize that social networking sites won't take the initiative to protect privacy, as we live in the real world walking on the streets. Users need to evaluate privacy risks when every time they upload information online. Since we cannot give up social networking online, this might be a more effective way to protect our privacy.

4 Public and Open to the Public[6]

Even though we agree with the common standpoint by courts, that the social networking site is a public sphere based on its function of communicating and networking, and we are exposing ourselves in public while uploading information onto Facebook, there is still a difference between information in public and making information open to the public. Information posted on Facebook is a kind of notification, a kind of states, and a kind of expression, in order for us to communicate with others and speak our thoughts out. Information in public only shows the meaning that the information subject doesn't mind people to know or may know his/her state or thoughts. There is no further authorization to anyone to

[6]In the USA, people discuss the right of publicity, defined as the right to control the commercial use of one's identity, in the field of intellectual property. But what I emphasized in this paragraph is that people reuse information found in public without notifying information subject or owner. This reuse includes transmission, distribution, reproduction, or collection as a data analysis.

use or reproduce the information. I am not building a special rule to explain the transmission of information. It is based on the social convention that we have to respect others. If we need the owner's agreement before taking something away, there is no difference when the object is information. Furthermore, thinking of the spatial connotation of privacy, in some situations privacy will be fringed at the very beginning when this private area is invaded, such as the confidential; but in most cases the problem arises because something is taken out of this private area, where we concern about more reproducing information and affecting our self-identity. Moreover, we are unable to evaluate information risks and lack of sufficient knowledge for perception. A general consent (e.g., agreement of Facebook privacy policy) cannot accurately reflect our true intention to dominate information in order to achieve self-actualization, but also cannot help us to avoid this known or unknown information privacy. Therefore the fact we expose our information in public will not absolutely lead to the conclusion that people can assess, reuse, and reproduce information without any condition.

5 Conclusion

Finally I concluded two arguments in this chapter. First, I observed the similarity between Facebook ticker and the sidewalk, and explained Facebook should be understood as the digital model of our society by this feature. Every user may be under others' watch in their daily life. To be completely anonymous online in our belief is partly imaginary. We imagine that hiding ourselves behind the screen, we still can watch others and their every move. Being invisible among people makes us feel safe. Unfortunately, that goal is not what the social networking sites are built for. The social networking site should be understood as our society digitalized. Thus, we have to understand that our digital activities are continuingly under someone's watch, just as our daily activities in public. Users have to realize this phenomenon while they are doing activities on social networking sites. We then can protect our privacy effectively and make ourselves feel safe, instead of feeling fear of no reason or being useless to keep ourselves away from the convenience of technology. Second, I emphasized that even if we do these digital activities in public, this does not of course mean we let our information open to the public. Whether this information or personal data can be accessed or used will depend upon how we define the privacy and its protection, discussed in the third part of this chapter. In short, if we don't want others arbitrarily use our information, at least personalizing our Facebook privacy settings is necessary.

References

1. Philosophy for Change: Foucault and Social Media: Life in a Virtual Panopticon., https://philosophyforchange.wordpress.com/2012/06/21/foucault-and-social-media-life-in-a-virtual-panopticon/. Last visited 18 Feb 2017
2. Rayner, T.: Why Facebook and Twitter are the virtual Panopticons of our time, http://www.abc.net.au/radionational/programs/philosopherszone/why-facebook-and-twitter-are-the-virtual-panopticons-of-our-time/5089494 Last visited 18 Feb 2017
3. D. Brin, *The Transparent Society: Will Technology Force Us to Choose between Privacy and Freedom?* (Perseus Press, Cambridge, MA, 1998)
4. Facebook Research., https://research.fb.com/three-and-a-half-degrees-of-separation/ Last visited 18 Feb 2017
5. L. Morales, Social media evidence: What you post or tweet can and will be used against you in court of law. The Advoc. (Texas). **60**, 32–43 (2015)
6. Facebook Ticker., https://www.facebook.com/help/255898821192992. Last visited 18 Feb 2017
7. Facebook Privacy Police., https://www.facebook.com/privacy/explanation (last visited 18 Feb. 2017)
8. Facebook adjusts controversial 'real name' policy in wake of criticism. https://www.theguardian.com/us-news/2015/dec/15/facebook-change-controversial-real-name-policy. Last visited 18 Feb 2017
9. H. Niedzviecki, *The Peep Diaries: How We're Learning to Love Watching Ourselves and Our Neighbors* (City Lights Publishers, San Francisco, 2009)
10. Foucault, M., *Discipline and Punish: The Birth of the Prison*, New Edition. Translated by Alan Sheridan (Penguin, New York, 1991)
11. Facebook Privacy Settings and Tools, https://www.facebook.com/settings?tab=privacy&section=composer&view. Last visited 18 Feb 2017
12. E.W. Sholl, Exhibit Facebook: The discoverability and admissibility of social media evidence. Tul. J. Tech. Intell. Prop **16**, 207 (2013)
13. M.-L. Wang, Information privacy: A contextualized analysis. (in Chinese) Chung Yuan Fin. Econ. Law Rev. **32**, 59–105 (2014)
14. Solove, D.J., *The Future of Reputation: Gossip, Rumor, and Privacy on the Internet* (Yale University Press, New Haven, 2008)
15. H. Nieenbaum, *Privacy in Context: Technology Policy and the Integrity of Social Life* (Stanford Law Books, California, 2009)
16. J.-R. Yeh, Towards the spatial connotations of privacy: Contextual articulation of the basic rights by the grand justices. (in Chinese) Acad. Sinica Law J. **18**, 1–40 (2016)

Identifying Influential Spreaders in Complex Networks with Probabilistic Links

Pavlos Basaras and Dimitrios Katsaros

Abstract Dynamic complex networks illustrate how "agents" interact by exchanging information in a constantly changing network. Typical examples of such networks are online social networks or human contacts. This article contemplates the common distribution of time that user-nodes spend on their activities, and describes a method for identifying real-time influential spreaders. We model the reciprocal activities of actor-nodes with probabilistic links and propose a technique for identifying influential spreaders in complex networks with probabilistic edges. The proposed measure, namely, *ranged Probabilistic Communication Area* ($rPCA$), is evaluated under the susceptible-infectious-removed (SIR) model, where the results illustrate that $rPCA$ can detect very effective spreaders in a networked environment with probabilistic edges.

1 Introduction

Real-world entities often interconnect with each other through explicit or implicit relationships, by transient and continuous ways to form a complex network. Social Networks (SNs) illustrate such complex interactions between individuals, and show how information, political views, frauds, advertisements, or rumors (*data*) flow through networked populations. Consider the most popular SNs (Facebook or Twitter), where users gain access to the Internet and their social activities through diverse wireless devices (smartphones, laptops, ipads) and become embedded to the Internet infrastructure swiftly for various and different time spans of their everyday lives, to interact, exchange opinions and ideas, or simply act like tuners for advertisements. Facebook self-reported statistics note that smartphone users check online 14 times a day, while an average user spends daily 40 min on the site. Now meditate on the vast amount of data traversing through such networks and how this

P. Basaras · D. Katsaros (✉)
University of Thessaly, Department of Electrical and Computer Engineering, Volos, Greece
e-mail: pabasara@inf.uth.gr; dkatsar@inf.uth.gr

T. Özyer et al. (eds.), *Social Networks and Surveillance for Society*, Lecture Notes in Social Networks, https://doi.org/10.1007/978-3-319-78256-0_4

magnitude of information has evolved through time. As reported in [1] in 2007 we had an average of 5000 tweets per day whereas in 2013 we were at 500 million tweets on a daily basis [2], representing a five orders of magnitude increase. From the above considerations one could argue on what share of these vast data is actually being 'seen' by its corresponding *audience*, that is, friends, followers or broadly speaking from the connected society, and on how this is further affected by the different time spans that individuals spend on their social activities.

It is evident that users cannot follow such immense traffic of data, but what of time-limited messages or alerts? As an example let's reminisce the *Twitter, Faster Than Earthquakes* event. On August 23, 2011, it took 30 s for an earthquake to travel from Washington DC to New York, but tweets were fast enough to reach NY quicker than half a minute. To account for many such cases for example of natural disasters, Twitter has launched the *Twitter Alerts: Critical information when you need it most* program in September 2013 for its users to receive reliable information during these times. In this study we emphasize on such *Real-Time Data, RTDs*, that need to be 'made known' to the largest possible portion of a social network at a short time interval (i.e., within a few minutes or hours) and on the fact that this particular info will serve no further purpose in larger time spans (e.g., days or weeks). Consider an enterprise announcing a discount of a certain 'hot product' but only for a limited stock or a limited time offer, aiming to attract large masses of consumers. A preeminent question arises; which users should be the targets for incentive that will initiate a cascade of informed-interested people and increase as much as possible the number of potential buyers?

Although we presented the problem in terms of activities over technological social networks, the issue of the effect of concurrent 'activity' is present in other types of complex networks as well, such as human contact networks and their relationship to infectious disease transmission. Theoretically, in such networks a short interaction between a susceptible and an infectious person could lead to a comparable amount of ingested infectious material as that of a long interaction assuming that the short interaction is more intensive than the long one. However, prolonged contacts tend to be more intensive than short contacts [3].

Motivation and Contributions

The issue of identifying influential spreaders in complex networks is a well-studied topic that received increased attention in recent years [4–6]. However, for this particular framework of data that we are addressing in the present study, the different patterns in the concurrent activities of 'connected' users will constitute the most essential ingredient for detecting the *Real-Time Influential Spreaders, RTISs*, rather than simply focusing in a static image of a social network and traditional approaches. At this point, we should note that both RTDs and RTISs are connotations to characterize data with relative short lifetimes and influential spreaders for such cases, respectively.

Empirical observations [7–9] note that users in SNs are not active around the clock, and they show a complex behavior and distribution over the time they spend on their social activities. A probabilistic framework that follows such complex behavior could portray the possibility of a link-connection to exist, that is, when connected users are active, and the dissemination process is in progress. A relative approach is that reported in [8] where the authors illustrate a probabilistic model that accounts for a node-user to be active or not (and thus his connections to be present or not) at the time for example of a disease outbreak or broadly speaking a diffusion process. It is thus an important feature that we need to consider in order to quantify the strength of the corresponding propagation.

In a similar approach to [8], we model the existence or absence of connections, rather than users, by annotating weights on links that correspond to the *mutual time* that connected users spent on their separate social activities. Intuitively if we could locate those nodes that are the starting points for paths of users which *share at a great degree common time in their online social activities*, it could provide valuable insights into better approximate the spreading capability of users and thus more efficiently 'control' the spreading process of RTDs. By conducting simulations and experiments in different Social Networks, we will see how the proposed identification technique, namely, *ranged Probabilistic Communication Area* ($rPCA$) effectively combined the activity schedules of connected users, identified the most influential spreaders and outperformed the competing techniques in various scenarios.

The present article discusses the issue of detecting influential nodes in complex networks with probabilistic links and makes the following contributions:

- Investigates the issue of detecting real-time influential spreaders by considering the mutual time connected users spend on their online social activities.
- Proposes an adjustable centrality measure, the range Probabilistic Communication Area ($rPCA$) that accounts for such characteristic and real- time data.
- Thoroughly evaluates this centrality measure under diverse competitive techniques in different real networks.

The rest of this article is organized as follows: An overview of relevant important works for the identification of influentials is presented in Sect. 2. Section 3 presents the proposed algorithm. In Sect. 4, we describe our experimental environment, competing techniques, and evaluation criteria. In Sect. 5 we evaluate the performance of the adversaries and finally in Sect. 6 the conclusions.

2 Related Work

The literature on the problems of maximizing the spread of influence and of identifying influential spreaders in complex networks is quite rich during the last decade. In this section, we only mention but a few among many important studies. We should also categorize networks depending on the pattern of their connectivity, that is, directed or undirected networks in order to discuss the direction of the propagation and finally emphasize on directed networks. The first problem was posed in [10] and later investigated further providing more efficient algorithms, for example, in [11–13]. Newer approaches to the design of centralities include concepts such as κ-path centrality [14] and distributed algorithms for identifying influentials based on random walks [15]. Other graph-theoretic methods include the k-shell decomposition of a network [6], where the authors contend that a node's location may be the determining factor that defines the influence potential of that node. Other approaches based on several shortcomings of k-shell are presented in [5, 16, 17], whereas local techniques that combine effectiveness and efficiency are proposed in [4, 18].

All these works concern single-layer complex networks. However, the last few years, we are witnessing an initiative in the analysis of new kinds of complex networks, where the interacting entities are assumed to belong to more than one network, called layers. Online social networks, financial systems, transportation networks are such networks to name a few; more detailed examples can be found in [19]. The study of spreading processes in multilayer networks has started to attract significant interest [20]. The works most closely related to the current article, that is, to influentials detection, are those reported in [21–24]. The blending of all layers into a single one and then application of traditional options for influentials detection are proposed in [21]. A generalization of the k-core is proposed in [22] but it results in a vector of values that cannot be used in a straightforward manner for detecting effective influential spreaders. In [23], the authors proposed an called KS, which follows the intuition of [25], that is, aggregates the shell indexes of its neighbors, and moreover combines the intra- and interlayer spreading rates. However, to our understanding, incorporating the unknown spreading rates, of (and between) the layers, is not realistic. In [24], very elegant methods based on tensor analysis are proposed.

Considering a directed social network, a user i is called a follower of j if there is a directed link from i to j ($i \rightarrow j$), namely, i can receive information from j. Thus for these network cases, the diffusion takes place through the incoming connections of a node-user. To detect the most influential spreaders in directed social networks, researchers often apply the $PageRank$ algorithm [26] where a node i is considered as influential if it is pointed by many other and important nodes. It is a random walk algorithm that was first used for ranking relative contents of web pages. A variation of PageRank, namely, $LeaderRank$, was proposed in [27] by introducing a ground node to the initial network, connected to all other nodes through a bidirectional link. LeaderRank identifies nodes which lead to quick and extensive spreading. On the other hand, LeaderRank is tolerant of spurious and missing links, which benefits applications with noisy data. In summary, LeaderRank outperformed PageRank not only by better identifying the influence potential of nodes, but also by converging faster to the final scores, and by being more robust to noise and spammers.

Via assigning degree-dependent weights onto links associated with the ground node, $weightedLeaderRank$ was presented in [28]. For this approach, the authors allow nodes with higher in-degree to get more scores from the ground node. Since the in-degree of a node directly indicates its influence, it is natural to weigh nodes according to their influence. Weighted LeaderRank outperformed its immediate predecessor by identifying more influential spreaders, by having higher tolerance to noisy data, and by having higher robustness to intentional attacks.

Finally, TwitterRank [29], also a variation of PageRank, was developed for identifying influential spreaders in Twitter. The fundamental difference of the two algorithms is that TwitterRank develops a topic-sensitive random walk, that is, the transition probability between users in Twitter is topic-dependent; in a way this generates a topic-sensitive network structure. Despite being better than PageRank, the design of TwitterRank takes into account a number of tweets a twitterer publishes, which makes it susceptible to manipulations if a twitterer deliberately publishes a large number of tweets.

As we mentioned earlier, users gain access to their networked environment through diverse wireless devices for arbitrary lengths of time and different frequencies. Such interacting behavior in social platforms resembles that of temporal networks. Quite often temporal networks are separated into two categories based on time sequences and time intervals for the interactions between connected individuals in communication networks. In our study, however, we are searching for connected individuals who have common online activity, that is, they do not necessarily exchange messages at arbitrary times but rather they are concurrently active at regular times. This can be considered as another simplification of temporal networks where we discuss the probability of existence of interacting paths based on such observations. For more details on temporal network analysis, readers are referred to [30] and references therein.

3 Proposed Technique

In this section, we present our proposal, the *range Probabilistic Communication Area* ($rPCA$).

Complex Networks with Probabilistic Links

A complex network $G(V, E, w)$ is a directed graph where V is the set of vertices (nodes), and E is the set of pairs of vertices (edges). Every edge is described by a weight $w \in [0, 1]$ and a direction. Each vertex involves in- and out-neighbors. As usual, the number of head endpoints adjacent to a user-node is called its *inDegree* (k_{in}), and the number of tail endpoints defines the node's *outDegree* (k_{out}). The weight values associated with every edge define a network structure which describes the probability for any two connected nodes to be both active, for example, during a diffusion process. As we will see later in our experimentation the mining and efficient use of such information will prove a valuable asset for the spreading of $RTDs$.

r-Hop User Communication Paths (UCPs)

A user communication path (UCP) on a directed complex network is a directed path consisting of *n* individuals and *n-1* connections among them such as no user appears more than once, for example, $a \to b \to e \to j$ in Fig. 1. For simplicity, the example network is a Directed Acyclic Graph (DAG). To complete our definition, we also need to define the range for such interacting paths as the number of connections that form it or the hop distance from the initial node, for example, a to j. For our technique, the communication paths emanating from each individual node will define its significance in the network. The weight values on the connections will be used to investigate on the quality of paths through which a user i "sees" the rest of the network in range or in other words *to search for users which share common time to their social activities*. Ideally we would like to identify user paths such as $a \to x \to z$, however, in a realistic networked environment, we cannot expect users to have identical online activity schedules. Furthermore since those node paths are probabilistic in nature, we need to also quantify the strength of those paths. Hence we apply the following formula to measure the strength of an r-hop interacting path ($SUCP^r$):

$$SUCP^r = \sum_{j=1}^{r-1} w_j \cdot w_{j+1} \tag{1}$$

where r defines the range of a particular UCP and w_j is the weight value at j hop distance from the originator, that is, the weight of the corresponding connection. Intuitively if we could rate nodes on the basis of their UCPs, we could set the right paths for the spreading of real-time data.

This methodology can sustain the case when we want to model topic similarity among nodes. In this case, each link connecting nodes i and j can have one more weight factor, namely, $\sigma_{i,j} \in [0, 1]$ that will describe the topic coherence among neighboring nodes [31]. Therefore, the total link strength will be evaluated as $w_{i,j} \times \sigma_{i,j}$. In this article, we assume $\sigma_{i,j} = 1 \ \forall i, j$.

Up to this point, we presented our proposal for quantifying the strength of a UCP. However, how to effectively combine the weight values associated with the corresponding connections in a communication path and define its significance is still an open issue. Another formula could be to simply acquire the product of its weights, however, such consideration will provide no distinction for paths with relatively equal-weight probabilities. For example in Fig. 1, for the interacting path $a \to b \to e \to i$, we would obtain a value of 0.063. The same value, however, would be attained if we sorted the weights in any possible way, for example, by reversing the probabilities of $b \to e$ and $e \to i$ or by placing the weakest interaction first and thus decreasing the probability of existence for the path. Another policy could be to assign a measure of importance for a specific weight depending on its hop distance from the originator, that is, weights closer to the initial node in a UCP are perceived as more vital. However, except for the fact that a tunable parameter

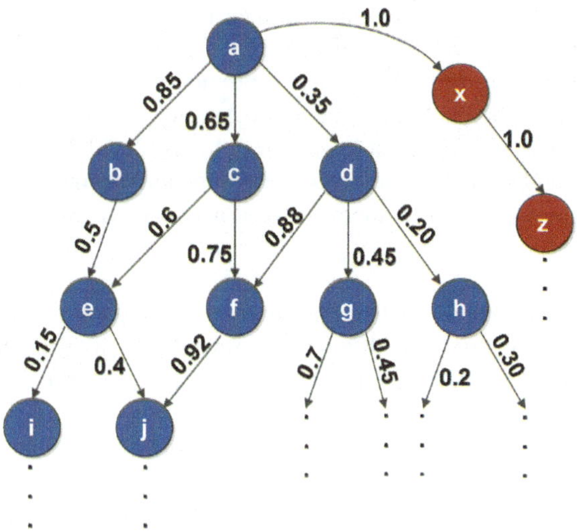

Fig. 1 $rPCP$ identifies nodes which possess the characteristic that from these nodes emanates "strong" paths. For 2 hops distance: $2PCA(a) = 17.283$ and $2PCA(b) = 1.1$ assuming that both i and j have 2 outgoing neighbors and x, z are hypothetic nodes, that is, not included

would have to be added, the significance of an interacting path like $a \to d \to f \to j$ which starts with a relatively weak weight and henceforth is composed of strongly connected users would be belittled with such consideration.

Range Probabilistic Communication Area (rPCA)

Following on these requirements, we built our proposal for defining centrality measures over graphs with probabilistic edges for range-limited neighborhoods. The $rPCA$ value of a node i within a specified range r is computed as the sum of $SUCP^r$'s emanating from i as follows:

$$rPCA(i) = \sum_{j=1}^{n} SUCP^r(j). \tag{2}$$

Note that nodes quite often share similar vicinities, that is, they may have a large number of common friends, and thus a certain path may be traversed by more than one way, for example, $a \to b \to e \to i$ and $a \to c \to e \to i$. For paths of interaction with hop distance greater than 2, the appearance of cycles, for example, $i \to j \to k \to j$ is a frequent phenomenon, especially when studying social networks often characterized as community networks, that is, dense connections within neighbors in the same community. However, considering "cycles of interaction" and thus returning to previous paths (or revisited node regions) are very likely to degrade an algorithm's performance, and thus these occasions are omitted by definition from our algorithm. In Fig. 1, we illustrate a toy example for the $rPCA$ method.

The proposed centrality measure can be *defined for both, the entire network* (*$*PCA$), and for neighborhoods around each node*. It is within our scope to maintain locality in order to provide an effective and efficient algorithm that can be applied in large-scale networks and real-time applications, and thus the range of $UCPs$ is limited at low values, that is, 2 and 3. Generally, we could search to any number of hops; however, we understand that increasing the range of $UCPs$ beyond the 90-percentile-diameter of a network (cf. Sect. 4) will provide little additional information to our approach since only 10% of the total size of the network is yet uncovered.

Although we have presented our method via the out-links of a node, when information flows through the in-neighbors of the network nodes as in our evaluation of Sect. 4, the implementation of UCPs is straightforward by following the in-links.

4 Performance Evaluation

For the evaluation purposes, we had to select appropriate competing methods, use networks with probabilistic edges, and also propagation models. In this section we describe our simulation environment and data sources.

Competing Techniques

In this subsection, we briefly describe the competing algorithms used in our simulation to evaluate the proposed method. Note that since information will flow through the in-neighbors of the network nodes the competitors are computed accordingly.

A diverse list of competitors were chosen regarding geodesics, the position of a node in the network, local techniques, or random walk-based approaches. A plethora of studies so far use the local degree centrality of a node to provide a baseline method for measuring the influence of nodes in complex networks.

(1) Likewise in our experimentation, we apply the weighted version of the approach. The *weighted degree centrality* ($wDeg$) of a node-user i or equivalently the strength of i is defined as the sum of the weights from the connections incident on i:

$$wDeg(i) = \sum_j w_{ji} \tag{3}$$

where j depicts the neighbors of i, that is, those nodes that i can exert influence, and w_{ij} stands for their associated weights.

(2) The farness of a user-node i is defined as the sum of its shortest distances to all other nodes of a network and the inverse of farness is noted as the closeness centrality of i. For its weighted implementation ($wClo$), the weights will describe how close or how far connected individuals are to each other as given by the formula:

$$wClo(i) = \sum_{j} \frac{1}{d_{ji}^{w}} \tag{4}$$

for all different j nodes of a network. In our framework, $wClo$ aggregates the weights on the shortest path and thus likewise our approach combines the weight values to provide an alternate technique that measures the strength and probability of existence for those paths.

(3) Shortest-path betweenness centrality describes the number of shortest paths that use a node i in order to reach other nodes of a network. Previous studies [5, 6, 18] found its performance insufficient to measure the influence potential of nodes in complex networks. Here we evaluate its performance in a relatively different experimental environment of weighted interactions and find similar conclusions ($wBet$):

$$wBet(i) = \sum_{s \neq i \neq t} \frac{\sigma_{st}^{w}(i)}{\sigma_{st}^{w}} \tag{5}$$

where σ_{st} is the total number of shortest paths from s to t and $\sigma_{st}(i)$ depicts the number of those paths that pass through i.

(4) A weighted version of the PageRank algorithm where the weights are proportional to the probabilities that a random walker will select a particular edge when choosing an outgoing connection from the current user-node [32]. Therefore, edges with larger weights are assumed to be traversed more frequently and are thus more important:

$$wPR_i(t+1) = (1-d) + d \cdot \sum_{j=1}^{N} \frac{w_{ji}}{\sum_{l=1}^{N} w_{jl}} wPR_j(t) \tag{6}$$

where w_{ji} is the probability of visiting node i from j if j is an in-neighbor to i otherwise $w_{ji} = 0$, d is the damping factor accounting for random jumps (in our experimentation, we assume no such occasions) and N stands for the total number of nodes in the network.

(5) The next and final competing algorithm weightedLeaderRank (wLR) was found to be more effective for the identification of influentials than PageRank and LeaderRank in directed networks [28]. Furthermore, it was proven more tolerant to noisy data by adding or removing links from the original network. We understand, however, that the traditional wLR algorithm does not use any

information regarding the weights of the links. It is used in our framework as a baseline method to measure the loss impact for not taking into consideration information through the weighted interaction. Nonetheless, our experimentation showed interesting results. As mentioned, it is a variant of LeaderRank, which introduces a ground node connected to all nodes of a network and recursively assigns scores to nodes depending on their k_{in}:

$$wLR_i(t+1) = \sum_{j=1}^{N+1} \frac{w_{ji}}{\sum_{l=1}^{N+1} w_{jl}} wLR_j(t) \tag{7}$$

where w_{ji} is equal to 1 if there is a directed link from j to i and 0 otherwise. If the destination node is the ground node then $w_{jg} = k_{in}^{\alpha}$ of j, where α is a free parameter set to 1 in our experimentation.

For the directed and weighted implementation for most of the above algorithms—excluding wLR and $wDeg$—we use the "igraph" R package.[1] igraph considers the weights assigned to each link as costs, that is, the largest the value the weaker the path. However, in our experimentation, weights indicate the strength of a link and thus we invert the original weight values for $wBet$ and $wClo$. A very popular method for the identification of influentials is the k-shell decomposition analysis [6] and its weighted versions, for example, [33]. However, to the best of our knowledge, there is no formal definition of the algorithm for directed and weighted networks. Could we have used measures such as μ–pci ? To such methods which are based on link counting and coreness, it is not clear how to quantize a "fractional degree" to its integer counterpart. Besides, such a conversion would loose significant part of the information carried by the probabilistic link.

Simulation Settings

Datasets

Nowadays there is a wealth of real datasets which concern complex networks; however, it is hard to find many input networks with probabilistic links with varying size and topology and varying distribution in the links' probabilities. Thus, in this article, we follow a dual methodology: We work with a real complex network to prove the applicability of our method in a real setting, and also use four real (initially unweighted) complex networks, which we annotate their links with probabilities drawn from various distributions, so as to test the scalability, effectiveness, and efficiency of the proposed method across a range of network sizes and link weight distributions.

[1] http://igraph.org/r/.

Table 1 Networks base attributes

Network	Nodes (V)	Links (E)	Diameter	90-EPD	E/V	Type
ego-Twitter	81,306	1,768,149	7	4,5	21.74	Social
soc-Slashdot0922	82,168	948,464	11	4,7	11.54	Social
soc-Epinions1	75,879	508,837	14	5	6.7	Social
wiki-Vote	7115	103,689	7	3,8	14.57	Social

The real probabilistic network is a contact network measured by the SocioPatterns collaboration[2] using wearable proximity sensors in a primary school, and covers 2 days of school activity. The sensors detect the face-to-face proximity relations (contacts) of 242 children [34]. The weight of a link is the aggregated contact duration of a pair of children. We normalize the links into the $[0, 1]$ interval by dividing each weight with the maximum weight found in the network. The experimental results which concern this real network are presented in Sect. 5.

The procedure for annotating the network links with weights is described in the following lines. We obtained our experimentation networks from the Stanford Network Analysis Platform [35]. For our evaluation purposes, the experimented networks were selected based on their connectivity, that is, three networks with a relatively equal number of nodes and decreasing in the number of their respective connections and finally a significantly smaller network. Specifically, we used the *ego-Twitter* network crawled from public sources, where followers receive information from their followees; *Soc-Epinions1*, a who-trust-whom social network of a general consumer review site, where users choose whether or not to trust reviews on products; *soc-Slashdot0922*, a technology-related news website, which allows users to tag each other as friends or foes; and finally *Wiki-Vote*, where nodes represent Wikipedia users and a directed edge from node i to node j represents that user i voted on user j. The base attributes of the aforesaid networks are listed in Table 1. The 90-effective-percentile-diameter (90-EPD) denotes the number of edges needed on average to reach 90% of all other nodes.

Generation of Probabilistic Links

For our simulation, the probabilities for the edge weights are assigned based on the Zipfian distribution for a range of skew values when the parameter $s \in [0.1, 0.9]$. The Zipfian distribution depicts the frequency of occurrence, for example, of a word randomly chosen from a text or the population rank of a city randomly chosen from a country. In our framework, it will depict the frequency of strong interactions. As s increases, we increase in the skewness for the distribution of weights and thus the strong weights will become more rare. In this study, we assume than any two connected nodes would share some common time of networked social activity, but

[2]http://www.sociopatterns.org.

also there are no identical schedules, that is, $w \in [0.1, 1)$. The resultant weight values will stand for the mutual time spent by nodes on their online social activities and thus depict the probability of an edge to be present or not at the time of the diffusion process. Links with values close to 1 are mostly active in our inspection time, whereas values near 0.1 are considered mainly inactive. According to these probabilities, we take ten 'snapshots' of the input graph resulting in ten abstract network images. Similar to [28] to obtain statistically unbiased results, we repeated the computation 100 times for each vertex in each network image, that is, averages over 1000 spreading processes.

Propagation Model and Influence

As far as the diffusion model is concerned, we employ the widely used susceptible-infectious-removed (SIR) model. SIR is commonly used for studying the spreading of epidemics in complex networks, where the infected nodes will either get immunity or die [36] and thus is suitable for our experimentation. We assume that an interested user propagates "data" only once, that is, users will not repeatedly send the same information to their respective vicinities. The Susceptible-Infectious-Susceptible (SIS) model is another popular method also used for the spreading of epidemics. SIS, however, has no immunity (like flu), and thus nodes get reinfected and further contribute in the diffusion. However, such consideration in our framework would include the provision of incentives to users in order to motivate them for propagating a certain datum a number of times.

In this study, we model the penetration of RTDs in a networked environment, with fixed transmissibility (infection rate) λ, for all user-nodes. SIR models three possible states:

- The susceptible state S, in which the S nodes are vulnerable to infection.
- The infected state I, in which the I nodes try to infect their susceptible neighbors and succeed with probability λ.
- The removed state R, in which nodes have recovered from infection and cannot be reinfected.

The diffusion proceeds as follows: In the initial phase, all nodes are in the S state except one node in I. An infected node is given a single chance to infect its susceptible neighbors and succeeds with probability λ. Immediately after and without loss of generality [28], the node enters the R state. The process continues until there are no nodes left in the infected state. Similar to [5] given a directed network, the influence of a node i, denoted by (IF_i), is defined as the average number of removed user-nodes at the end of the spreading process if i was the initially infected node. Conventional techniques for measuring the *epidemic thresholds* [37] in the evaluated networks cannot be employed in our case study, due to the probabilistic nature of in- and out-neighbors, and thus we confined our work to a range of λ values between 1 and 10%.

Evaluation Criteria

Kendall's Correlation (τ)

To evaluate the ranking abilities of each competing method with respect to the actual spreading potential of each node, we use the Kendall's Tau 'b' rank correlation coefficient (τ) [38]. It is a statistic used to measure the association between two measured quantities, for example, (*2PCA*, *IF*). When $\tau = 1$, we have a perfect correlation, indicating that when node i is ranked before j by some method, that is, with greater *2PCA*, then its spreading capability is also higher. For $\tau = 0$, the measured entities are considered neutral, whereas $\tau = -1$ implies opposite correlation. Generally, the closer we get to 1, the better the correlation of the evaluated approach.

Fraction of Ranked Nodes: False Index

As depicted in Fig. 2 for the lower spreading rates, there is a large number of users with zero influence, for example, over 70% for the soc-Slashdot0922 network when $\lambda = 2$. Applying Kendall's correlation to such unfiltered values will provide harsh results. In our experimentation, we take a closer look for each λ value to provide a more complete assessment and thus the ranked sample used for the ranking process will be composed of user-nodes with $IF > 0$, namely, p users. To complete the evaluation of the results and conclude on which technique better identifies the influence power of nodes, we also need to provide an assessment for the rest of the

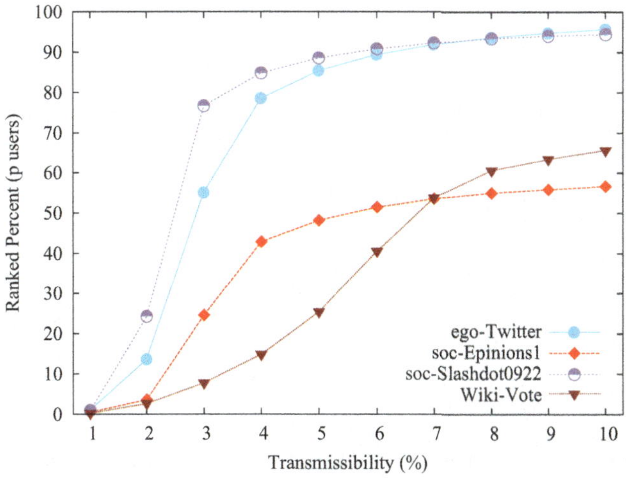

Fig. 2 Ranked percent with respect to the total number of nodes of each network case for all evaluated λ values, that is, nodes with $IF > 0$

$1-p$ non-ranked users. The *False Index* depicted in Figs. 3(upper right) to 4(bottom right) fills this void. To obtain the False Index, we calculate for each node in $1-p$ the number of nodes in p whose index is lower from that particular node's. In other words, we measure the average number of nodes which, although did not succeed in propagating, were ranked with higher index by some users in p, for example, with greater $2PCA$. Reasonably, a small False Index indicates better results.

5 Results

Impact of Infection Probability

In this section, we evaluate the efficiency of each competing method in ranking nodes according to their actual spreading potential, when varying in the strength of the propagation in four different Social Networks. For the distribution of links in Figs. 2, 3, 4, 5, 6, 7, s is set at 0.7. In almost all the evaluated networks, we observe that the most abrupt changes in the curves of correlation for all methods occur at

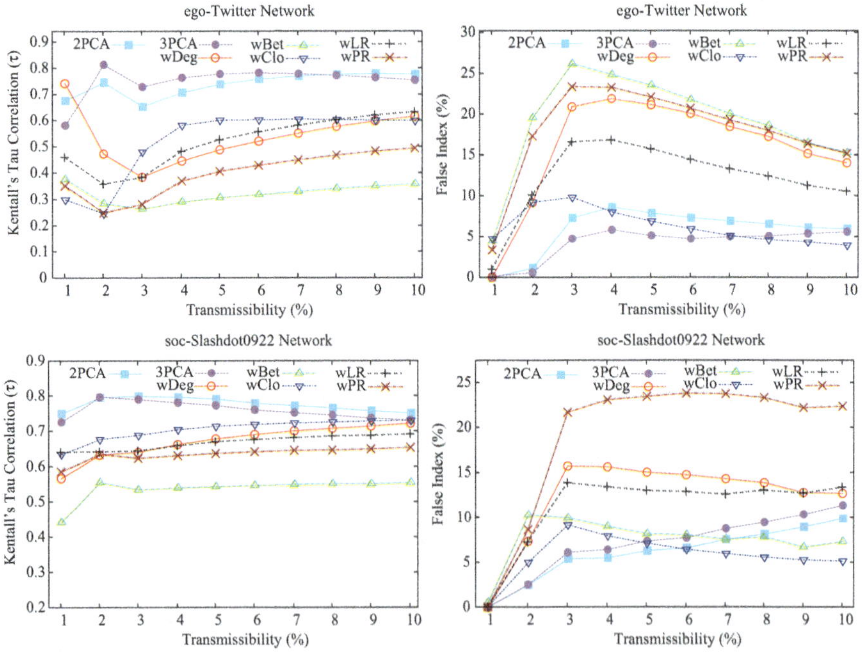

Fig. 3 (Top: Twitter plots). In almost all different spreading rates for the ego-Twitter network, the proposed technique significantly outperforms its competitors. (Bottom: Slashdot plots). For the soc-Slashdot0922 network, we observe that our approach coincides with the rest of the competing algorithms only for the higher spreading rates

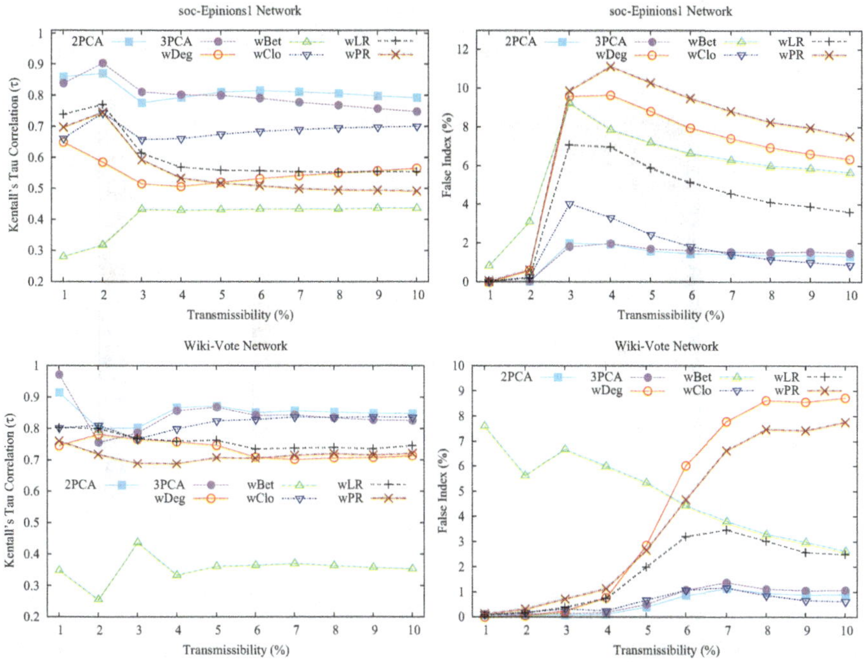

Fig. 4 (Top: Epinions plots). As the spreading rate increases, our two-fold approach maintains its superior performance as compared to the rest of the competing techniques. (Bottom: Wiki plots). For the final network case, an oscillation for the most accurate ranking is observed at the lower spreading rates. Nonetheless, the proposed technique is found within the higher τ values

the lower λ values. This is partly because the largest leaps in the percent of the ranked p users occur within the fist few increments of the spreading rate, that is, when $\lambda < 4$ for most of the evaluated networks (about 6 for the Wiki-Vote), where we observe that the fraction of ranked nodes drastically changes. For instance as illustrated in Fig. 2, for the Twitter network when $\lambda = 2$ the p nodes constitute about 15% of the total size of the network, whereas when we move to $\lambda = 3$ this percent is close to 58%. The changes in the curves of τ however are not only due to the increasing number of the p users used in the ranking process. As the spreading rate increases, the influence of nodes from previous λ values also changes and the same may happen to the ranking between those nodes in subsequent spreading rates.

Considering the results in Fig. 3(upper left), Kendall's coefficient for 2–3PCA when $\lambda = 2$ is above 0.75, whereas the rest of the competing techniques are found below 0.5. Similar observations can be made for the soc-Slashdot0922 network, that is, the largest differences in τ are found at the low spreading rates. For Fig. 4(top left) and (bottom left), however, the above observation does not hold. For these cases, we observe a more sedate behavior of the curves as we increase in λ. Apparently, the probabilistic property of the networks affects the dynamics of a cascade and thus in Sect. 5 we investigate on the quality of the probabilistic links.

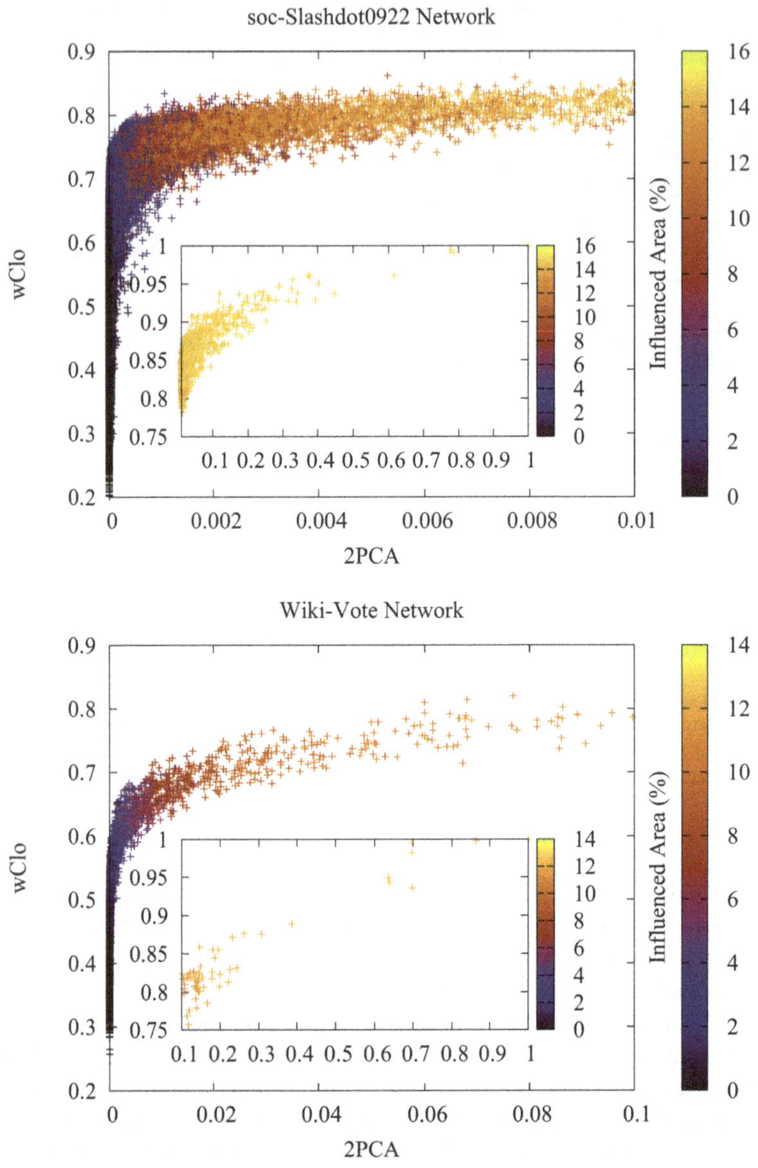

Fig. 5 *wClo* was found to coincide with the proposed technique in a few configurations. The presented heat plots illustrate that influence is closely related to 2PCA. On the contrary, for *wClo*, we observe that the medium values depict an amplitude of influence values

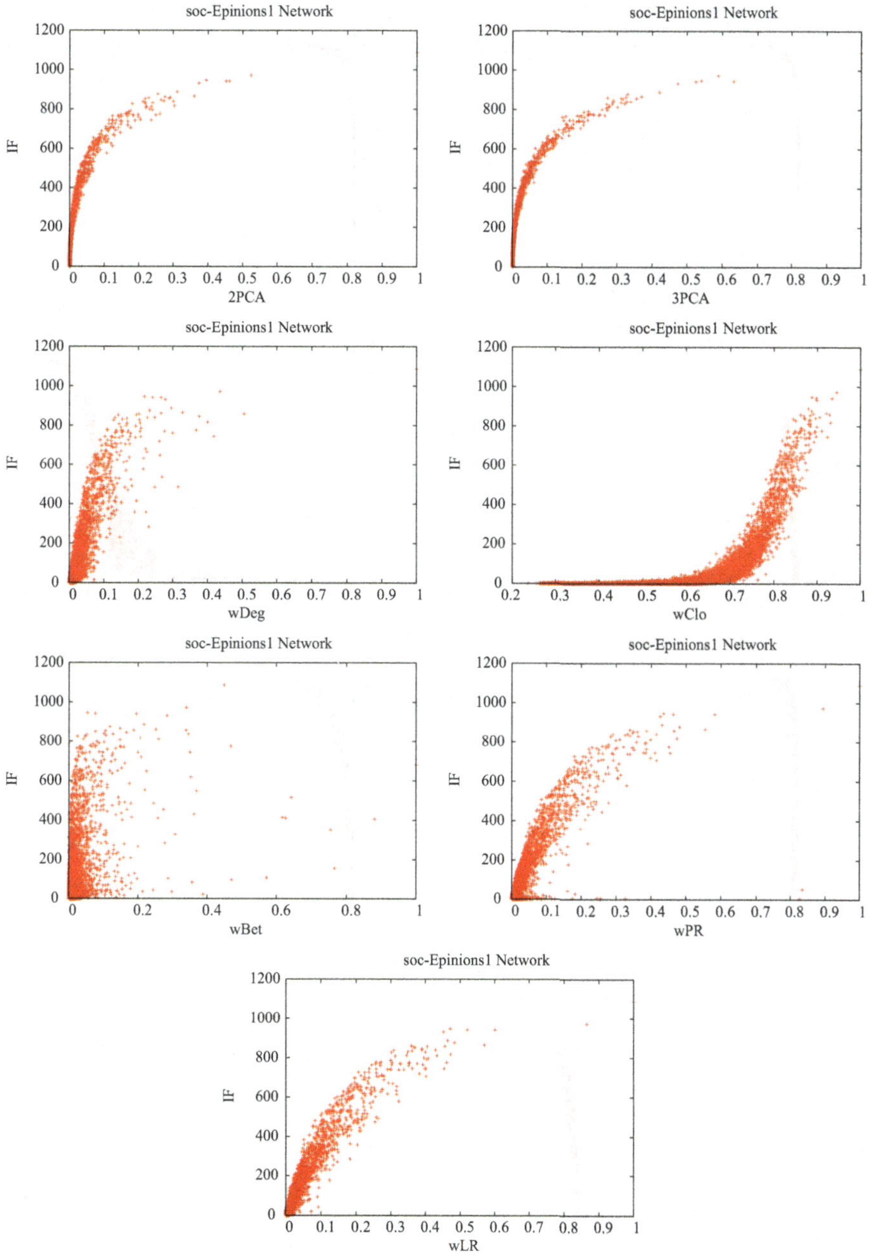

Fig. 6 Spreading rate is set at 3%

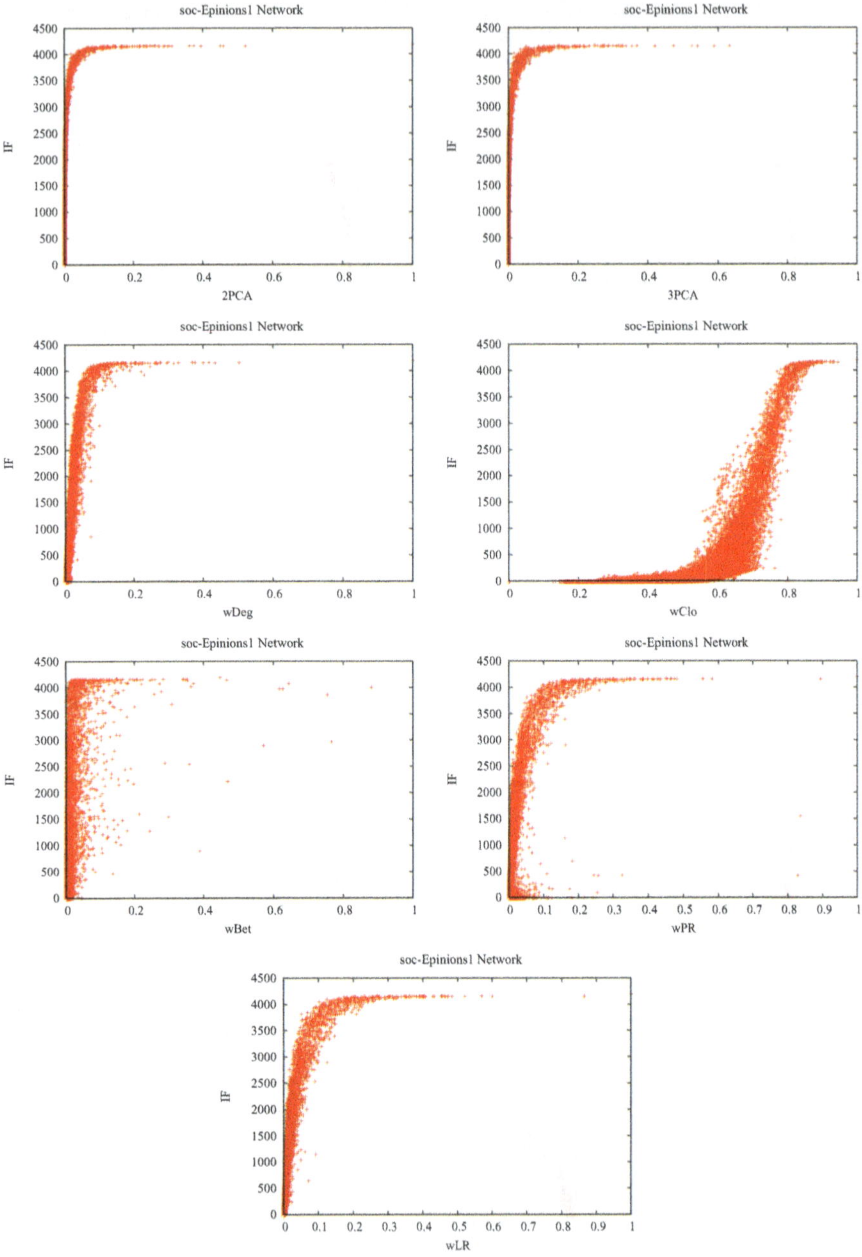

Fig. 7 Spreading rate is set at 6%

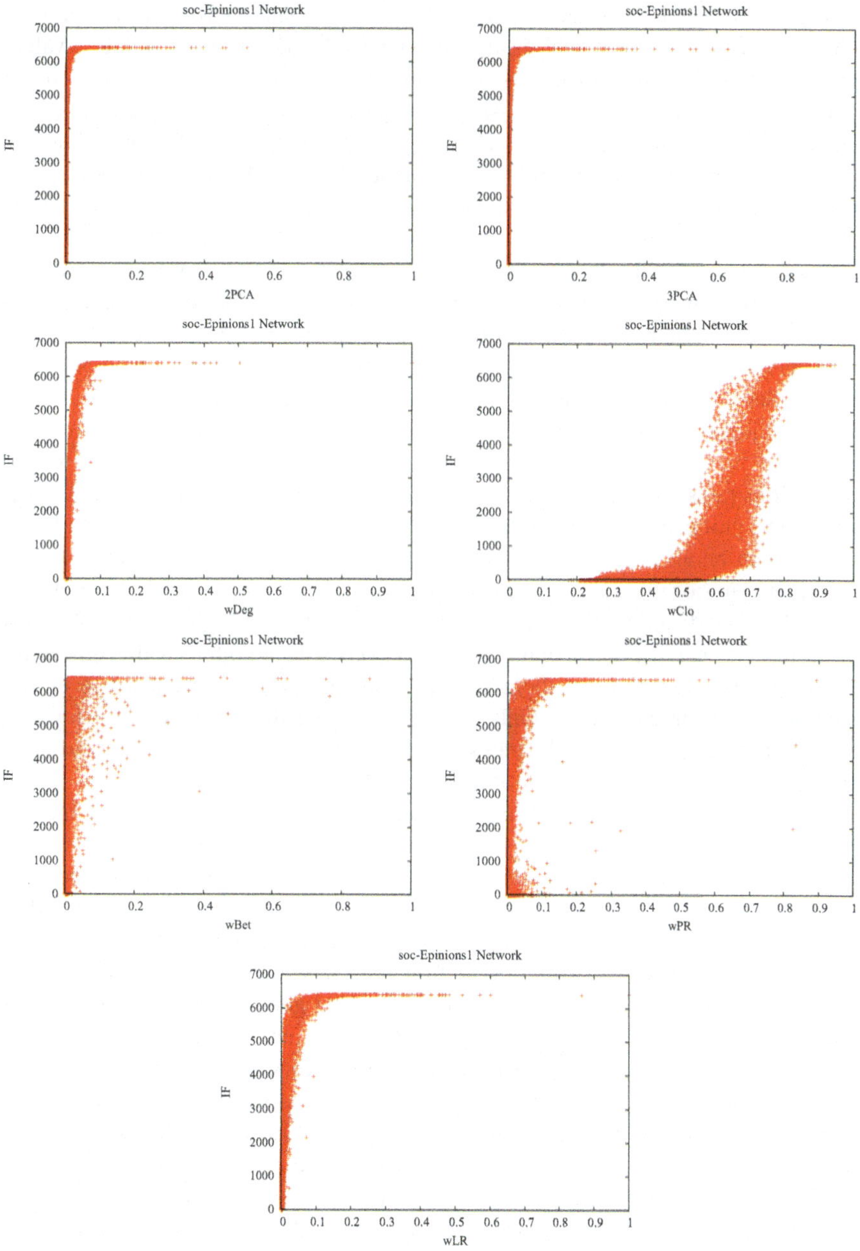

Fig. 8 Spreading rate is set at 9%

In Fig. 3(bottom plots), when λ is around 9%, $wDeg$ and $wClo$ coincide with our approach. It should be emphasized that for very large values of λ, the τ values of correlation for the competitors are bound to cross over and oscillate. This is due to the fact that on such occasions an epidemic will occur regardless of the characteristics of the originator. For the higher spreading rates, the true influential nodes are very likely to get infected at some point as the diffusion progresses, and thus result in an epidemic outbreak, even though the originator is not truly an influential. Besides by using large λ values, the role of individual nodes in the diffusion process will no longer bear significance [5, 6, 16, 17]. When considering the different ranges of our approach, we can see that for the low spreading rates there is an oscillation for the most accurate ranking between the two methods. However, as we increase in λ for all network cases $2PCA$ always obtain higher τ values. Such consideration indicates that local information of a node's surroundings (communication paths) is more favorable as we increase in the spreading rate.

For an overview on the False Index, 2–$3PCA$ is found at the lower percentages. $wClo$ illustrates similar behavior; however, the rest of the competing techniques illustrate significantly higher values. Note that the False Index does not provide any information about how accurate the ranking for the p nodes is but rather acts as a further criterion for each respective technique. Ideally, we would obtain a zero False Index indicating that none of the $1 - p$ users has higher index than any node in p. Generally, a low False Index coupled with a high τ will promote the most efficient algorithm for the addressed issue. Clearly, the proposed technique supports the desired outcome. Only at the higher spreading rates in Fig. 3(top left) and (bottom left) 2–$3PCA$ illustrates higher False Index. However, these cases are trivial since as depicted in Fig. 2, almost all nodes are within the p node set.

Focusing separately on each competitor, $wDeg$ is used as a baseline method to illustrate how complete locality serves in quantifying the spreading power of a node. For its overall evaluation, it is indicated by our results as a moderate approach for our ranking purposes and real-time data. When considering its False Index, we can see that $wDeg$ is rated among the three worst-performing methods in all evaluated network cases. This observation indicates that simply considering the total strength of a node's local connections is not a good indicator to quantify its spreading influence. For example, a high $wDeg$ index may be accumulated by many but otherwise weak interactions which in our framework is interpreted as regularly absent connections. To our perception, such occasions will result in insignificant influence results, and may be the reason for $wDeg$'s high False Index values. As another contributing factor to its medium performance, we can say that $wDeg$ does not 'carry' any information about the position of the node in a network. Therefore, although a node might me connected to its immediate vicinity with strong links, if it is positioned in the periphery of a network [6], reasonably we expect that its influence will be rather diminished.

Another interesting point seen through our simulation is the performance of wLR. Reminisce that for this particular method no information about the activity schedules is used, and thus we expected a relatively low correlation in our framework of weighted interactions. Nonetheless, it was proved rather compensatory as a competing technique which indicates that wLR may indeed be a good indicator for the spreading potential of nodes in unweighted networks. Although the τ values of correlation for wLR are significantly lower from our approach, it was found to be comparable and even better on many occasions when considering the rest of the evaluated techniques, for example, as illustrated in Fig. 3(bottom left) or 4(top left). Generally, its performance can be considered relatively similar to $wDeg$'s; however, we can conclude that wLR provides a more accurate ranking if we consider the False Index of the two aforementioned techniques, that is, $Wdeg$'s False Index is always higher.

In contrast to wLR, wPR accommodates information from the weighted interactions in the sense that links with higher weights are traversed more often. Both techniques were found to follow approximately the same trend in all evaluated networks as the spreading rate increases, that is, their illustrated curves either both ascend or descend. However, our experimentation showed that wLR obtained higher correlation with influence and also significantly lower False Index values. Nodes with no outgoing links, the sink nodes, which are indeed present in the evaluated networks are not well handled by PageRank, since they decrease the PageRank overall [39]. To our understanding, such inefficacy overestimates the spreading power of a node and may be the reason for wPR's low correlation and the highest False Index values. Generally, through such methods users pointed by many other and important users are elected as strong influencers; however, as also noted in [26, 40] quite often the k_{in} of a node is not sufficient to characterize its influence capacity.

Next we investigate on $wBet$ and find that this particular method has the worst performance in all evaluated networks, while other studies [5, 6, 18] also note its inability to capture a node's influence capacity. Its low efficiency can be explained if we consider that through $wBet$, node-users who are unique intermediates for some other nodes (or medians leading to different communities) are elected as important entities. However, in such cases their capability for influence and propagation may well be overestimated if these nodes lead to regions with sparsely connected nodes or small sized communities. In our simulation where the problem of identifying influential spreaders is further enhanced by considering the time distribution of nodes social activities, $wBet$ will be at a further disadvantage if those links correspond to nodes with highly uncommon time spans. As a final observation for $wBet$ in our experimentation, we found that among the p user-nodes, there was a significant amount of nodes with zero betweenness scores, which also explains the high False Index values of the competitor. This observation indicates that nodes which do not reside in any shortest path may be more influential from nodes with higher betweenness scores, and further confirms that the influence cannot be measured through the shortest paths that pass through a node.

Finally, $wClo$ utilizes useful data through the weighted interactions, in the sense that nodes connected through weak links are considered to be relatively far to each other. However, as shown in our simulation in most of the illustrated results, simply aggregating the strength of the connections to obtain the average distance of a node to the remaining nodes of the network lacks when compared to our approach.

We attribute its lower performance to the following: first, although the effective diameter for all network cases is relatively small, for example, between 4.5 and 5, there are still more than 8000 nodes for Twitter and Slashdot0922 networks, and more than 7500 for soc-Epinions within a diameter of 7,11, and 14 hops, respectively. However, considering long interacting paths would include a mixed set of connections, that is, a relatively long path may be composed of both strong and weak links. To this end, we expect that techniques that utilize global information of a network's connections to define the significance of a node in the network will furnish varying results. Figures 3 to 4 confirm our statement. Lastly, unlike our approach, $wClo$ considers a single communication path to all other nodes from the focal node, and in particular the shortest (strongest) paths to those nodes. Nonetheless, rather than a single strong path, it may be more favorable to take into account a number of interacting paths that reach a single user-node, that is, multiple paths, in our framework of complex networks with probabilistic links.

In Figs. 3(bottom plots) and 4(bottom plots), we found that $wClo$ coincides with our approach significantly and thus we advance to thoroughly understand the relation of the two methods, that is, $2PCA$ and $wClo$ with influence in Fig. 5. The spreading rate is set at 10% for both networks where the aforementioned techniques are closer. The heat values depict the IF in percent, for each user-node with pair values ($2PCA$,$wClo$). For nodes with the same pair values, the average IF is used. Note that each axis is normalized to its largest corresponding index. Moreover the outer plots are ranged up to a certain value of $2PCA$ which is then resumed in the embedded charts of each corresponding network to illustrate more precise results. From these figures, we can further argue that $2PCA$ is the better indicator for the spreading influence of nodes in complex networks with probabilistic links and further strengthen the superiority of the proposed technique. From the embedded charts, we can understand that the highest index values for both methods indeed correspond to the most influential spreaders. However, from the outer plots, for example in soc-Slashdot09022, we can see that for a range of values in 0.7 to 0.8 for $wClo$ there is a wide variety of influence scores, that is, approximately between 4 and 14%. Such observation indicates that the medium values of the competitor cannot distinguish the influence potential of nodes in contrast to $2PCA$ which provides a more accurate ranking. We found similar conclusions when comparing $wClo$ to $3PCA$.

Overall, our experimentation showed that for our technique, paths limited in the near neighborhood of the focal node, that is, two-hop UCPs, are usually sufficient to characterize its role in an epidemic. In our framework, the probabilistic property of the networks affects the diffusion dynamics and thus we urge for a technique that effectively handles the different probabilities for connected nodes. Our ranged approach was found quite effective and efficient that better identified influential spreaders in most of the observed network scenarios.

Spreadability

In this section, we illustrate the results in Figs. 6, 7, 8 where the x-axis represents the values of each competing technique and the y-axis the corresponding influence (IF). For a better overview on the competitors, we illustrate the results for a range of λ values, that is, 3,6 and 9%, respectively. We found similar qualitative results for the rest of the evaluated networks. In this section, we measure the different influences of user-nodes with approximately equal index scores, that is, for relatively equal $2PCA$s what is the range of IF values. Apparently the smaller the amplitude of IF the better the technique. It is evident that among all competing methods $2PCA$ illustrates the best correlation with influence which is indicated by its thin ascending curve which increases in IF as $2PCA$ increases. Similar conclusions can be made for $3PCA$; however, as we increase in the spreading rate $2PCA$ illustrates better performance. These observations are consistent with the results in Fig. 4(top left). For wLR and $wDeg$ as previously noted, we found similar behavior. For the lower spreading rate ($\lambda = 3$), we observe that there is a wider range of IF values for $wDeg$ and moreover the slope of the curve for wLR is greater. These observations explain the higher τ for the latter in Fig. 4(top left). Nonetheless, as the spreading rate increases, their difference diminishes and when $\lambda = 9$, $wDeg$ has a thinner curve and thus better correlation.

When $\lambda = 3$, for wPR we notice that there are a few nodes with very large values, for example, around 0.8, which have insignificant or zero influence. Moreover in 0.1 to 0.3 the amplitude of IF ranges from almost zero influence up to the largest value. These observations persist as we increase in the spreading rate; however, such inefficacies will significantly affect the correlation of the technique with influence. A similar amplitude of IF values is illustrated for $wBet$, although for significantly more user-nodes, which further distances the algorithm for use in the identification of influentials. Finally, following on the performance of $wClo$, we understand that the lower values of the technique correspond to users with low or insignificant IF, whereas the highest index scores indicate the most influential nodes. The problem of the competitor lies to its medium values, that is, for a range around 0.7 (which increases as λ increases) where we observe a wide variety of influence scores, quite similar to Fig. 5(upper) for soc-Slashdot0922 network. Such observation afflicts the competence of the algorithm for the addressed issue and further strengthens the superiority of the proposed technique.

For any influential spreader detection algorithm in order to be characterized as an efficient one, it is important to have a steep ascending curve, which is 'thin', especially as we move to larger values of a technique along the x-axis. 2-$3PCA$ was found to adopt such behavior, taking the lead on its competitors in both steep upward slope and smaller deviation in IF in all network cases.

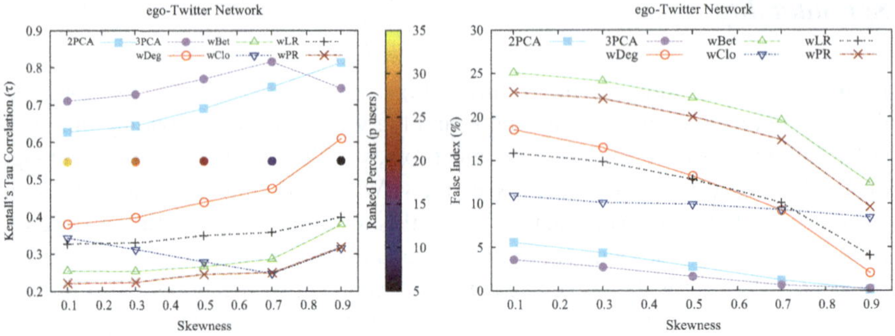

Fig. 9 Ranging in skewness for the distribution of links (spreading rate is set at 2%)

Impact of Zipfian Skewness

In this set of experiments, we investigate on the skewness of the Zipfian distribution as s increases. Due to similar results, we present only those for the ego-Twitter network in Fig. 9. The spreading rate is set at 2%. The percentage of nodes that succeeded in propagating (p users) is illustrated with the colored cycles mapped to the corresponding heat values in the palette. As a first observation, we note that as we increase in s for the distribution of links, the number of users that are able to propagate in their respective vicinities decreases. This phenomenon is anticipated as we distance our experiments from uniform distribution and gradually force the weights toward the lower possible values. In our framework, such configuration results into frequently absent connections resembling a realistic social environment where we cannot expect node-users to have largely common time spans for their social activities.

As shown in Fig. 9 most of the competing techniques illustrate similar behavior in both evaluation criteria, that is, decreasing and increasing trend for the False Index and τ, respectively. For the lower s values, we observe only small increases in τ. However, as we further increase in s, the changes in τ become more evident. This is due to the fact that for the larger skews, the now fewer strong links and interacting paths become more clear for the competitors. This remark is most visible when $s > 0.7$ where we observe the most significant changes for all methods. wLR, however, shows minor changes in τ, an observation somehow coherent with [28] where the authors explain the robustness of the technique in "noisy" networks, that is, missing links.

When we have a fairly good distribution for the weights (low skewness), we observe that $3PCA$ obtains the highest correlation followed by $2PCA$, whereas the rest of the competing techniques obtain significantly lower values in τ. This observation indicates that when we have many strong interactions, that is, nodes with highly common activities, accumulating information from relatively long UCPs indeed results in better correlation. In an opposite scenario where node-users have significantly different schedules (large skewness), the strong weights become more rare. Using long paths composed of weak interactions will degrade our algorithms performance which explains the steep fall of $3PCA$ for the higher s values. Conversely thinking, we can understand the illustrated behavior of $2PCA$ which uses short-ranged communication paths and takes the edge on our ranged approach in the aforesaid cases. The significant difference in the False Index values between the competitors and 2–$3PCA$ further strengthens the superiority of our method. For instance, 2–$3PCA$'s "misjudgment" near 0.9 becomes almost zero, whereas in most of the evaluated scenarios (different skews) it is found below 5%. Finally we conclude that in a framework with probabilistic links that portray the property of active nodes as described in our work, considering multiple paths and moreover multiple alternative paths (unlike $wClo$) is a first step for devising an appropriate method for the identification of real-time influential nodes.

Evaluation with a Real Complex Network

After the detailed performance evaluation of the methods across a range of network sizes and link weight distributions, we use a real weighted complex network in order to confirm the practicality of the problem examined and also to further support the superiority of the proposed method. Recall from Sect. 4 that this is a contact network measured by the SocioPatterns collaboration[3] in a primary school. The sensors detect the face-to-face proximity relations (contacts) of 242 children [34]. The resulting network has 242 nodes and 4024 links, after removing the nodes terms as "Teachers" and their interactions, because the network offers no possibility to differentiate between different teachers. Figure 10 depicts the number of interactions

Fig. 10 Distributions of the link weight (i.e., aggregated contact duration) of the real weighted network

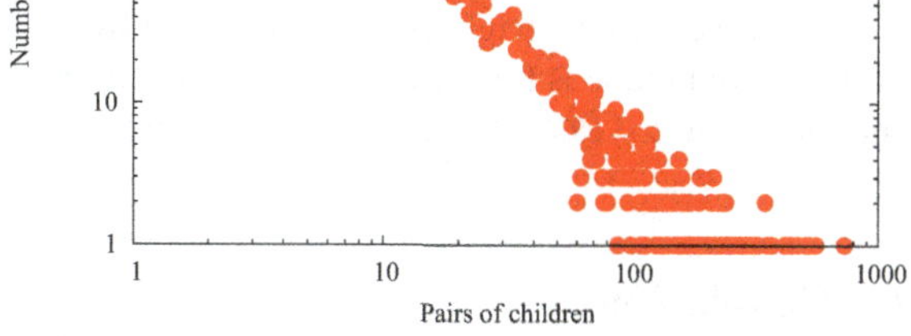

[3] http://www.sociopatterns.org/2015/01/a-high-resolution-social-network-measured-in-a-primary-school/.

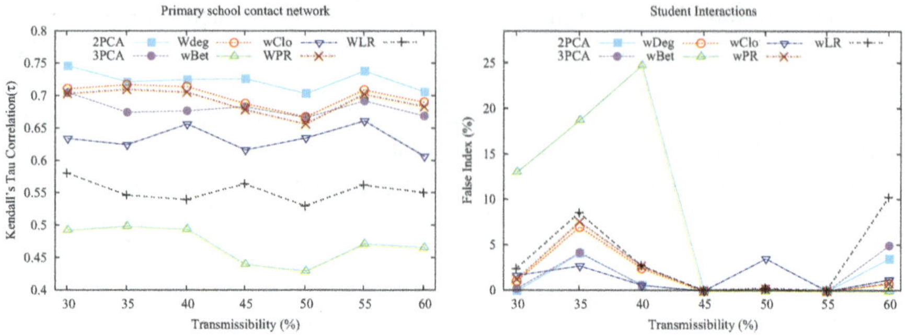

Fig. 11 Evaluation of competing algorithms over the real weighted network

per pair of children. According to the methodology of data collection (sensor beaconing), each contact lasts for 20 s. Thus, this figure shows in an equivalent way the aggregated contact duration of a pair of children, which is the link weight in our case. Evidently, this distribution follows a power-law, where the majority of the pairs of children have less than ten contacts.

The evaluation of the competing algorithms is presented in Fig. 11. The first comment concerns the transmissibility rates in order to achieve high enough infection. The generic comment is that the infrequent student interactions require higher transmissibility rates for successful transitions. Specifically, for the lower λ value, only about 2% of the network is infected, for example, from an emerging flu originating from the most influential student, whereas when $\lambda = 60$, the infected students rise up to 30%.

Regarding the performance of methods, we observe that the best strategy—consistent with our previous results—is $2PCA$, whereas $wBet$ is the worse strategy. The position of the second best-performing strategy is now occupied by $3PCA$, $wDeg$, and WPR (subject to some variation). The interesting thing is that $wClo$, which was steadily the third winner in our earlier finding, now is fifth. Based on the rankings we obtained for this real network and the conclusions by Figs. 10 and 9, we can say that the link weight distribution of this network is highly skewed for which networks we already have seen that the performance of $3PCA$ and $wClo$ degrades significantly. Finally, complementary to the False Index illustrated for the artificial networks, we observe (right plot Fig. 11) no different qualitative results, that is, the proposed technique is found at the lower false values, which further strengthens the superiority of 2–$3PCA$ for the addressed issue.

6 Conclusions

The evolution of social networks to date indicates that the amount of information flowing though user interactions is only going to increase. In this article, we argued on what portion of information remains 'unseen' from interested users due the continuous flow of data in such networks. With this consideration, we focus on 'pieces' of information with limited life spans, that is, for data that are interesting to some users but only for a limited time (RTDs). In order to push information into a network and spread RTDs to the largest possible extent, we need to account for users who share at a great degree common time in their social activities. With this demand, social networks must be remodeled to probabilistic structures. In this study, we used probabilistic links to simulate the probability of connected users with common social activity, and proposed a centrality metric, namely, $rPCA$, which accounts for probabilistic communication paths around the focal node. The proposed technique was evaluated under different spreading rates and distribution for the weight probabilities, and proved superior from its competitors in ranking nodes according to their true spreading potential. Finally, to our understanding, how each method uses-filters the lower weight values is a determinant factor to its performance, since users with low common time spans will contribute little to each other's influence. Moreover in order for RTDs to be substantially propagated, we need not only consider the strength of each individual link separately but rather as combined attributes within the interacting paths. For our future direction, we intend to apply different approaches for quantifying the strength of the UCPs and further improve our formula for the identification of influential spreaders. Also, other factors could be considered in defining the weights on edges, for example, the characteristics of the individual nodes, or the characteristics of the communities, could play a significant role in communication.

Acknowledgements This work was supported by the project "REDUCTION: Reducing Environmental Footprint based on Multi-Modal Fleet management System for Eco-Routing and Driver Behaviour Adaptation", funded by the EU.ICT program, Challenge ICT-2011.7.

References

1. K. Weil, Measuring tweets (2010). Twitter Official Blog, February 22
2. R. Krikorian, New tweets per second record, and how! (2013). Twitter Official Blog, August 16
3. T. Smieszek, Theoretical Biology and Medical Modelling **6**, 2–15 (2009)
4. P. Basaras, D. Katsaros, L. Tassiulas, IEEE Comput. Mag. **46**(4), 26 (2013)
5. B. Joonhyun, K. Sangwook, Physica A **395**(1), 549 (2014)
6. M. Kitsak, L.K. Gallos, S. Havlin, F. Liljeros, L. Muchnik, H.E. Stanley, H.A. Makse, Nat. Phys. **6**, 888 (2010)
7. A.L. Barabasi, Nature **435**(6), 207 (2005)
8. J.B. Holthoefer, S. Meloni, B. Goncalves, Y. Moreno, J. Stat. Phys. **151**, 383 (2013)
9. J.B. Holthoefer, A. Rivero, Y. Moreno, Phys. Rev. E **85**, 066123:1 (2012)
10. P. Domingos, M. Richardson, *Proceedings of the ACM SIGKDD International Conference on Knowledge Discovery and Data Mining (KDD)* (2002), pp. 57–66
11. D. Kempe, J. Kleinberg, E. Tardos, *Proceedings of the ACM SIGKDD International Conference on Knowledge Discovery and Data Mining (KDD)* (2003), pp. 137–146
12. J. Leskovec, A. Krause, C. Guestrin, C. Faloutsos, J.M. van Briesen, N.S. Glance, *Proceedings of the ACM SIGKDD International Conference on Knowledge Discovery and Data Mining (KDD)* (2007), pp. 420–429

13. C.T. Li, T.T. Kuo, C.T. Ho, S.C. Hong, W.S. Lin, S.D. Lin, Soc. Netw. Anal. Min. **3**(3), 341 (2013)
14. N. Kourtellis, T. Alahakoon, R. Simha, A. Iamnitchi, R. Tripathi, Soc. Netw. Anal. Min. **3**(4), 899 (2013)
15. B. Han, A. Srinivasan, *Proceedings of the ACM International Symposium on Mobile Ad Hoc Networking and Computing (MOBIHOC)* (2012), pp. 5–14
16. J.G. Liu, Z.M. Ren, Q. Guo, Physica A **392**(18), 4154 (2013)
17. A. Zeng, C.J. Zhang, Phys. Lett. A **377**(14), 1031 (2013)
18. D. Chen, L. Lu, M.S. Shang, Y.C. Zhang, T. Zhou, Physica A **391**(4), 1777 (2012)
19. S. Boccaletti, G. Bianconi, R. Criado, C.I. del Genio, J. Gomez-Gardenes, M. Romance, I. Sendina-Nadal, Z. Wang, M. Zanin, Phys. Rep. **544**, 1 (2014)
20. M. Salehi, R. Sharma, M. Marzolla, M. Magnani, P. Siyari, D. Montesi, IEEE Trans. Netw. Sci. Eng. **2**(2), 65 (2015)
21. M.A. Al-garadi, K.D. Varathan, S.D. Ravana, E. Ahmed, V. Chang, J. Intell. Fuzzy Syst. **31**(5), 2721 (2016)
22. N. Azimi-Tafreshi, J. Gomez-Gardenes, S.N. Dorogovtsev, Phys. Rev. E **90**(3), 032816 (2014)
23. Z. Dawei, L. Lixiang, L. Shudong, H. Yujia, Y. Yixian, Phys. Scr. **89**(1), 015203 (2014)
24. M. De Domenico, A. Solé-Ribalta, E. Omodei, S. Gómez, A. Arenas, Nat. Commun. **6**, 6868 (2015)
25. B. Joonhyun, K. Sangwook, Physica A **395**, 549 (2014)
26. S. Brin, L. Page, Comput. Netw. ISDN Syst. **30**(1–7), 107 (1998)
27. L. Lu, Y.C. Zhang, C.H. Yeung, T. Zhou, PLoS ONE **6**, 0021202:1 (2011)
28. Q. Li, T. Zhou, L. Lv, D. Chen, Physica A **404**, 47 (2014)
29. J. Weng, E.P. Lim, J. Jang, Q. He, *Proceedings of the ACM International Conference on Web Search and Data Mining (WSDM)* (2010), pp. 261–270
30. P. Holme, Proc. IEEE **102**(12), 1922 (2014)
31. F. Menczer, Proc. Natl. Acad. Sci. **99**(22), 14014 (2002)
32. R. Baeza-Yates, E. Davis, *Proceedings of the ACM International World Wide Web Conference (WWW)* (2004), pp. 328–329
33. M. Eidsaa, E. Almaas, Phys. Rev. E **88**, 062819:1 (2013)
34. V. Gemmetto, C. Barrat, A. Cattuto, BMC Infect. Dis. **14**, 694:1 (2014)
35. J. Leskovec, A. Krevl, SNAP datasets: Stanford large network dataset collection (2014). http://snap.stanford.edu/data
36. B.A. Prakash, D. Chakrabarti, N.C. Valler, M. Faloutsos, C. Faloutos, Knowl. Inf. Syst. **33**(3), 549 (2012)
37. L. Cong, H. Wang, P.V. Mieghem, Phys. Rev. E **88**, 062802:1 (2013)
38. M. Kendall, Biometrika **30**, 81 (1938)
39. P. Devi, A. Gupta, A. Dixit, Int. J. Adv. Res. Comput. Commun. Eng. **3**(2), 5749 (2014)
40. J.K. Kleinberg, J. ACM **46**(5), 604 (1999)

Decentralized Online Social Network Architectures

Francisco de Asís López-Fuentes

Abstract In recent years, online social networks have become very popular. Online social networks (OSN) have been useful to find former classmates or to improve our interaction with friends. Social networks are also used to establish new relationships with other users, organize events with friends, promote personal work, share multimedia contents, publish comments, or broadcast news. In this way, a huge amount of information is generated and consumed by millions of people via these networks every day. Although the benefits introduced by social networks are well known, these networks still have many challenges. Today popular online social networks are centralized and based on the client-server paradigm. Centralized servers are responsible for managing and storing all information. However, centralized structure introduces several drawbacks in the online social networks such as storage dependence, scalability, security threats, and privacy. P2P infrastructure have emerged as an alternative platform to deploy decentralized online social networks. However, P2P systems are limited by network performance, topology, or vulnerability to security attacks. In this chapter, we review P2P infrastructure as an alternative platform for developing decentralized online social networks. This chapter introduces the benefits of decentralized online social networks and studies its architectures in terms of performance and security.

1 Introduction

Online social networks (OSN) have become more and more popular every day, and increased the social interactions between groups and individuals. Many persons use these networks to find former classmates or interact with friends. Social networks allow establishing new relationships with other users, organizing events with friends, promoting personal work, sharing multimedia contents, publishing

F. de Asís López-Fuentes (✉)
Departamento de Tecnologías de la Información, Universidad Autónoma Metropolitana Cuajimalpa (UAM-C), Mexico City, Mexico
e-mail: flopez@correo.cua.uam.mx

T. Özyer et al. (eds.), *Social Networks and Surveillance for Society*, Lecture Notes in Social Networks, https://doi.org/10.1007/978-3-319-78256-0_5

comments, or broadcasting news. However users must be careful with their personal information published in these networks. Today most popular OSN sites provide free storage for users, who can share different contents such as photos, videos, and news. Most of these popular online social networks are based on centralized structures that are controlled by a single authority. A centralized approach in the OSN may introduce several drawbacks mainly related with security and privacy [1, 2]. For example, user's content dissemination can be used for business purpose or social data abuse [3–5]. Decentralization of the backend infrastructure has been proposed to deal with these limitations in the current online social network services. A decentralized online social network (DOSN) is an online social network implemented on a distributed computing platform, such as networked servers or a peer-to-peer (P2P) network [6]. In this context, a DOSN combines social and decentralized elements. On the other hand, P2P networks have emerged as a promising distributed information management platform. A P2P network is a distributed network formed by a group of nodes, which build a network abstraction on top of the physical network, known as an overlay network. In a P2P system, each peer can work both as a server and as a client at the same time. In a P2P network, the peers provide all available resources such as processing, memory, and bandwidth [7]. Therefore, when a new peer arrives to the P2P system the demand is increased, but the overall capacity too. This is not possible in a system based on the client-server model. Under this approach, different characteristics of a P2P network such as decentralized control, ad hoc deployment, or difficulty taking down can be some relevant reasons for the DOSN. In a decentralized online social network users keep control of their data and can use the social network locally, without Internet access. Although P2P networks offer a promising alternative for DOSN, P2P networks still have different challenges to be solved. An important challenge in P2P networks is security, because P2P infrastructure are not originally designed to withstand adversarial attacks. Due to its distributed nature, P2P networks are subjected to more intricate attacks compared to those inflicted on client-server networks. To evaluate a decentralized online social network (DOSN) are used different criteria such as architecture, type of service, development application programming interface, availability, scalability, privacy control, and business model [8]. The objective of this chapter is to share with a generic audience the advantages of the decentralized architectures for deploying online social networks, with a particular focus on the P2P paradigm. To this end, the chapter describes P2P infrastructure, and compares a pure distributed architecture against a hybrid architecture in terms of performance and security. The chapter aims to study how the performance and security of a DOSN are affected by the type of P2P architecture used. Specifically, the Kademlia and BitTorrent P2P protocols for a specific DOSN called Tribler are studied. The behavior of the Kademlia and BitTorrent schemes against harmful attacks such as Sybil and spoofing is also reviewed. The chapter also briefly reviews some video streaming systems.

The rest of this chapter is organized as follows. Section 2 introduces a background about decentralized P2P infrastructure. Section 3 presents related work about decentralized online social networks. A brief overview about Sybil and spoofing attacks is introduced in Sect. 4. Section 5 presents some results for Kademlia and BitTorrent related to their performance in the presence of security attacks. Chapter concludes in Sect. 6.

2 Decentralized P2P Architectures

A decentralized online social network is an online social network deployed on a distributed computing platform, which allows it to combine social and decentralized elements. Authors of DOSN approaches indicate that decentralization of OSN can be achieved using P2P technology or dedicated server using federation protocols [9]. This chapter focuses on the DOSN based on P2P networks, which have several properties common to P2P approaches such as scalability and robustness that can be considered as advantages for DOSN. However, DOSNs based on P2P networks are different from other P2P services mainly with respect to the updating of files and the privacy of the information. That is to say that in a social network the information is updated frequently, and files only are readable by peers that are specifically allowed to access them [6].

In a DOSN, the number of participant nodes, the storage capacity of nodes, and the efficiency of the network play a very important role, and they are crucial parameters to offer an efficient service [10]. In an architecture, different components of a system can be organized to achieve the system's goals. A DOSN architecture introduces an abstraction of the DOSN design space [9]. A general DOSN architecture can be represented by a three-layer system. Top layer represents extensions such as graphical user interfaces, and the applications. Intermediate layer is called the core which contains the main parts of the DOSN such as the access control, profile storage, and an overlay component to organize the communication between nodes (e.g., P2P overlay management). Finally, the lowest layer represents the communication network used by all participating entities [9]. In a DOSN two system components play a particularly important role: control and storage. Control consists of lookup services (user and content lookup) and identity management service, while storage consists of storage of user's content and ensuring its high availability. These components can be organized closely that resembles that of the P2P systems, which are classified as unstructured, semi-structured, and structured. This classification is based on how the nodes in the overlay structure are connected to each other.

Unstructured P2P Architectures

An unstructured P2P network is formed when the logical links among participating nodes are established randomly. In these systems, no user maintains an index, and system operations are usually carried out using flooding or gossip-like communication between users. Unstructured systems are designed more specifically

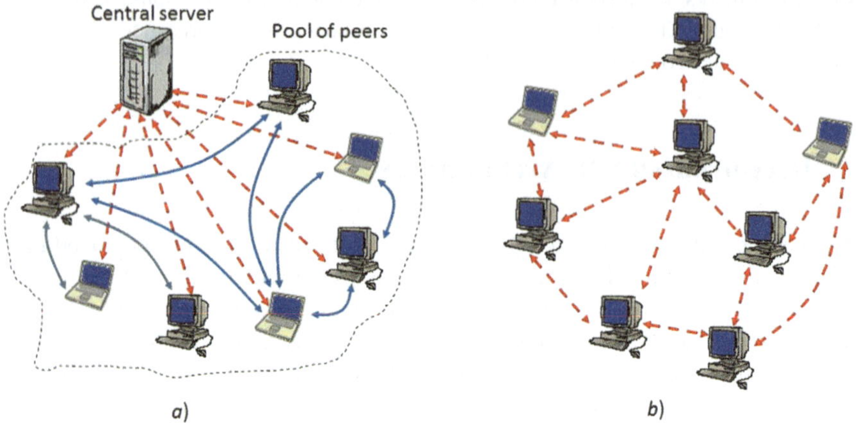

Fig. 1 Different unstructured P2P architectures. (**a**) Centralized P2P, (**b**) pure P2P [12]

for heterogeneous and distributed environments [8, 11] where maintaining strict restrictions on control data placement and network topology is not possible. Unstructured networks generate a high amount of traffic because the queries have to be flooded through the network. Most unstructured P2P systems are considered as the first generation of P2P systems, which were used to share and to store files. Unstructured systems can be further divided into [11] centralized P2P and pure P2P. Figure 1 shows the architectures for these different unstructured P2P approaches.

Semi-structured P2P Structures

A semi-structured P2P network is formed by a subset of all the users in the system (superpeers), which take responsibility for storing the index and managing other users. The superpeers are responsible for providing the interface to the rest of the users for carrying out different system operations. User participation for providing superpeer service can be voluntary or incentive based [8]. Figure 2 shows a semi-structured P2P architecture.

BitTorrent

BitTorrent is a protocol designed for exchange of large P2P files on Internet. In a BitTorrent network, a central server serves as core and manages bandwidth resources, routing, and communication between nodes without knowing the identity of each node and without storing any information. Thus, the server does not share files of any kind to any node and reduces the overload impact generated by distributing large files. In a BitTorrent network several basic computers can collaborate to efficiently

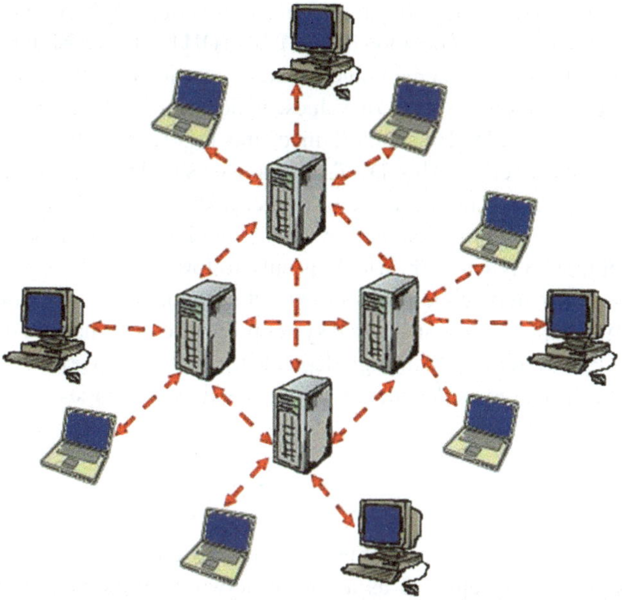

Fig. 2 Semi-structured P2P architecture (superpeers) [12]

distribute files to many requesting clients. A user uploads a file by creating a small torrent descriptor file and distributes it using conventional mechanisms as Web or e-mail. Several actors and components such as peers, leechers, seeders, trackers, and swarm form a BitTorrent network. All users connected to the BitTorrent network are called peers. In a BitTorrent scheme there are two types of peers: seeders and leechers [13, 14]. Seeders are users with a file without considering if they are the original creator or they just have published it. On the other hand, users that download files are called leechers. A tracker is responsible for the communication between peers, and keeps a global registry of all the downloaders and seeds of the corresponding file [15]. Initially, clients need to communicate with the tracker to identify peers from where they can download a file. Once started downloading the file, communication service between peers can continue without a tracker. However, clients can continue to communicate with the tracker in order to obtain statistics from new peers. Finally, all peers who share a torrent form a swarm. Swarming allows tolerating massive flash crowds for popular content [14].

Structured P2P Architectures

In a structured P2P network users directly participate to form a structured overlay or use services provided by a third-party structured overlay. Any query in the system can be resolved using the structured overlay in a bounded number of steps [8]. These systems are structured because they maintain a close coupling between the network topology and the location of data via a hash table (DHT). This hash table is used to precisely define the data placement and lookup operations. In these systems, each peer acts as a server for a subset of values, which can be data items or pointers to where data is available [16]. A DHT mechanism is responsible to handle peers joining/leaving the overlay. This DHT mechanism should be fully distributed in order to avoid a single point of failure or bottlenecks. A routing table with links to a small subset of peers is maintained in each peer. Then, these links can be used to construct the overlay network, which usually follows a regular topology such as ring, tree, mesh, or hypercube [16]. Structured systems show superior scalability and reliability compared to unstructured systems. On the other hand, a disadvantage of structured P2P systems is their high dependence on the network dynamic, which hinders to maintain a required structure for efficiently routing messages.

Kademlia

Kademlia is a distributed hash table protocol designed for decentralized P2P networks. Kademlia is deployed as a virtual network on an existing LAN/WAN network or Internet and its topology is based on the XOR metric. This metric is used to calculate the distance between points in the key space [17]. Several benefits are derived from this novel approach such as participants can receive lookup queries exactly the same distribution of nodes hosted in their routing tables or they can send a query to any node within an interval. These facts allow the participants to select routes based on latency or even send parallel asynchronous queries. Compared with previous P2P protocols, Kademlia's nodes send a reduced number of configuration messages to learn about each other. Collected knowledge is used to provide flexible routing via low-latency paths. In Kademlia, the timeout delays generated from failed nodes are avoided by using parallel asynchronous queries. Each Kademlia's node is identified by a 160-bit node ID. Kademlia protocol consists of the following RPCs: PING, STORE, FIND NODE, and FIND VALUE. These procedures allow to specify the network structure, and regulate communication between nodes and exchange of information. The communication between nodes is realized using UDP protocol.

3 DOSN Approaches Based on P2P Networks

Decentralized online social networks (DOSN) have lately received a lot of research attention and many kinds of DOSNs based on the P2P paradigm have been proposed by several authors. Some of these approaches to implement general-purpose decentralized online social networks are Diaspora [18, 19], PeerSON [20], Cachet [21], Prometheus [22], and Tribler [1]. In this section are described some more important characteristics of these approaches.

Diaspora

Diaspora is an OSN launched in November 2010 [18, 19], and it has been designed for real-world Internet scale and to preserve the user privacy. Diaspora is a network formed by independent and federated servers, which are managed by the individual users. The users' profiles are hosted in these federated servers; however, users can decide for themselves their preferred servers to store their information. In Diaspora, some users can decide to have a total control of their data by maintaining their own servers, while others might choose to join an existing server. Diaspora network uses posts as a way to conduce most communication between users. Diaspora manages two levels of privacy for the posts: public and limited. A public post means that it is visible to anyone, while a limited post means that it is only shared to a specific group of users. In Diaspora, users can organize their contacts into groups in order to share with them contents about a specific aspect. In such way, a user can choose only those posts from certain users that result from their interest. Diaspora server uses push design to send data and notifications to other servers, while the Diaspora network distributes data replicas to multiple servers [19]. This solution gives to Diaspora network a decentralized approach, but the local servers handle the privacy and security of the posts where they are stored, and the unencrypted user data could be exposed to read and write access. During a communication between users of different servers, the data are pushed and stored in both servers. Therefore, a trust communication is required.

PeerSON

PeerSON is an online social network (OSN) proposed by Buchegger et al. [20] that addresses three limitations in the online social networks: privacy issues, authentication, and Internet connectivity for all transactions. To solve these problems the authors use encryption and access control coupled with a peer-to-peer approach. In a PeerSON, peers need not necessarily be connected to the Internet to make use of their social network. A peer can restrict its communication to other PeerSON-enabled devices, but profiles of other users cannot be accessed. The various features provided by OSNs can broadly be classified into the categories of social link establishment, digital personal spaces, and means of communication [20]. These categories are represented in PeerSON. To reach this goal, PeerSON has a two-tiered architecture. Logically, one tier serves as a lookup service. The second tier consists of peers and contains the user data, such as user profiles. PeerSON currently uses a distributed hash table (DHT) for the lookup service. However, this service cannot assure informational self-service and nodes could derive certain information about a user by issuing recurring requests for the same hash. PeerSON assumes the existence of a public key infrastructure (PKI) to control privacy and identity management. In this way, users encrypt their content with the public key and distribute the key to the intended audience. Only those users who have the right key can access the contents [8], which allows the users to have fine-grained control over the access rights of their contents.

Cachet

Cachet is proposed in [21] as a purely decentralized OSN where users collaborate with each other to store their contents without any centralized service. Cachet has been implemented on top of DECENT [23], which is a modular architecture design for online social networks. DECENT employs a distributed hash table to store user data and features cryptographic protections. These features allow this online social network to offer confidentiality, integrity, flexibility attribute polices, as well as a fast revocation. In this way, user's data and their social relationship are not visible to unauthorized users. Data availability is based on data replication and authentication of updates. Cachet uses a distributed hash table as a base storage layer, but adds a gossip-based social caching algorithm that dramatically increases performance [21]. Online social contacts are used to immediately propagate new updates, retrieve cached updates, or discover other online contacts. Cachet combines decentralization, attribute-based encryption, and use of caches to provide high availability, low latencies, and flexible policies for protecting data. This OSN uses container objects to store data such as photos, status updates, or references to other containers. Containers are protected by cryptographic structures in order to ensure confidentiality while supporting multi-principal interactions without revealing policies or user relationships to the storage nodes. However, the encryption technique requires costly decryption operations of several contents from a user's social contact during the generation of its news feed [8]. Response time in Cachet is improved with social caching, which is maintained by a user using continuous secure connections with other nodes (online users).

Prometheus

Prometheus is introduced in [22] as a P2P service that collects and manages social information from multiple sources that are refereed as social sensors. Collected information is exposed through interface in order to implement nontrivial social inferences. This network employs user-defined policies to control the access to social data. Social sensors are user applications used to collect information about interaction with other users via different devices or the Web. Prometheus processs this collected information in order to create a decentralized, directed, and labeled multi-edged graph, which reports users who are connected and how they interact in social network.

Tribler

Another online social network based on P2P paradigm is Tribler [1], which is a research project developed in the Delft University of Technology and Vrije Universiteit Amsterdam. Tribler runs over BitTorrent and exploits social phenomena to content discovery and download cost. Tribler is a free software with a defined architecture and is currently available for Windows, Linux, and Mac. Based on these findings, we select this social network as the online social network to be evaluated on two different P2P infrastructure. Performance of both infrastructure is evaluated in terms of download time. Tribler has been proposed and implemented on top of the BitTorrent protocol in order to gain decentralization, availability, integrity, cooperation, and network transparency. To carry out this implementation the authors have modified and extended the existing BitTorrent client software. Tribler architecture involves different concepts such as social groups, megacaches, taste buddy, collaborative download, list of friends, geo-localization, recommendation, and peer similarity. Many of these components are integrated together via a user interface, which is a key and critical part of Tribler. User interface facilitates the formation of social groups. Bootstrapping is a task to find other peers in a P2P system. In BitTorrent this task is realized by a tracker. However, to solve this problem, Tribler uses superpeers and overlay swarm through an epidemic protocol called Buddycast. Superpeers are contacted by a Tribler peer to obtain an initial list of other peers in the system, while overlay swarm is used for content and peer discovery. Social phenomenon concept in Tribler has been implemented in the social networking module, which is responsible for storing and providing information related to social groups such as the group members and used IP numbers. In Tribler, megacaches are used as local storage to store each piece of context information received by a peer that is relevant to it based on its interest and tastes. This information is exchanged within social groups using the Buddycast protocol. As we have previously indicated, Tribler uses the BitTorrent protocol for downloading files. However, to achieve a significant increase in file download speed the BitTorrent Engine uses a collaborative downloader module which exploits idle upload capacity of online friends [1]. This collaborative module is based on a protocol called 2Fast [24], which helps to improve some limitations in the original BitTorrent protocol. In 2Fast, a peer can participate in a collaborative download taking one of the two roles: collector and helpers. A collector peer is interested in obtaining a complete copy of a particular file, while a helper peer can provide assistance to a collector peer during a file download. Both peers work similar as in the BitTorrent protocol; however, a collector peer can optimize its download performance by dynamically selecting the best available data source from a set of helper peers [1]. In this protocol, a helper peer works under an altruist approach, that is to say, a helper peer collaborates without requesting anything in return.

Other examples of OSN can be found in the literature such as Vegas [3] and Safebook [25], which means that decentralized online social networks are still an interesting research topic for academic and industrial sectors. This chapter reviews the simulation of an online social network for two different P2P infrastructure. To reach this goal an OSN called Tribler is selected as experimental social network to be implemented. On the other hand, Kademlia [17] and BitTorrent [26] are selected as the P2P architectures over which Tribler has been deployed.

4 Privacy and Security

OSNs suffer from various security and privacy problems, and distributed infrastructure have been suggested by several authors as an alternative to face these problems [9]. However, security issues also represent a great research challenge in the P2P infrastructure. Due to its distributed nature, the P2P networks are subject to more intricate attacks than client-server networks. To improve privacy in DOSN two general approaches have been suggested: access control list (ACL) and data encryption. An ACL allows that only authorized user can access or modify a specific content, while data encryption allows that only users who have decryption key can interpret it [9]. However several authors recommend to implement both approaches to provide greater protection to the content in the DOSN. On the other hand, there are several adversary models to attack privacy and security in the P2P networks. In this chapter we review two typical attacks called Sybil and spoofing attacks and how these attacks affect the P2P infrastructure.

Sybil Attack

Sybil attack is highly damaging for P2P systems, because a malicious entity creates multiple false identities of itself and uses them to influence the system behavior [38, 39]. A system affected by the Sybil attack can provide false information to other legitimate nodes, and create a majority of colluding malicious nodes in the overlay network. Sybil attacks are difficult to detect. Sybil defense schemes for online social networks are presented in [28, 29], which also helps us understand the characteristics of real-world social networks [29]. Protection strategies against Sybil attacks are analyzed in various publications such as [27, 30]. It is important to know how the performance of P2P structures for social networks is impacted by Sybil attacks. The structure of the social network plays an important role because if attacker knows this then he/she can launch a more efficient Sybil attack.

Spoofing Attack

In a spoofing attack, an attacker forges its identity to masquerade as another device, or even creates multiple illegitimate identities [31, 32]. Spoofing attack is a serious threat because it represents a form of identity compromise and can facilitate a variety of attacks such as steal data, spread malware, or traffic injection [31]. Due to this it is important to detect the presence of spoofing for its eliminations from the network. There are several different types of spoofing attacks; however, most common attacks are IP spoofing, ARP spoofing, and DNS server spoofing. In this chapter, our spoofing simulation is based on IP spoofing. An important part of the Internet is vulnerable to this type of spoofing attack [33–35]. IP spoofing is the act of forging the source addresses in IP packets. In this case, an attacker can hide its identity and location by masquerading as a different host. Duan et al. [35] state that IP spoofing will remain popular for the following reasons:

- IP spoofing makes it harder to isolate attack traffic from legitimate traffic.
- IP spoofing allows that packets with spoofed source addresses may appear to be from all around the Internet, which presents the attacker with an easy way to insert a level of indirection. So, a substantial effort is required to localize the source of the attack traffic.
- IP spoofing is used by many popular attacks such as man-in-the-middle attacks, TCP SYN flood attacks, and reflector-based attacks.

5 Comparison of P2P Infrastructure for DOSN

For the purpose of this chapter, we show comparative results of two kids of P2P for a DOSN called Tribler. The results of these experiments have been reported in [36, 37]. Tribler is deployed in two different P2P schemes: Kademlia and BitTorrent. Kademlia is a pure P2P infrastructure while BitTorrent is a hybrid infrastructure based on superpeer. These two kids of P2P infrastructure are evaluated against Sybil and spoofing attacks. Experiments have been done using 2500 nodes and downloading a file of 100 MB. In both scenarios, 15% are seeder nodes and 85% are leecher nodes. Initial reputation is random. A node is removed from system when its reputation is 0. Initial iteration starts with 2200 nodes because nodes with faults or with low reputation are removed in these experiments.

Performance of both P2P infrastructure impacted by the presence of noncooperative peers is shown in Fig. 3. Results show that Tribler based on BitTorrent can download the file faster than Tribler based on Kademlia, with equal number of non-cooperating nodes. This is because as time passes Tribler based on BitTorrent eliminates noncooperative nodes, which improves the system performance. Although the protocol ignores uncooperative nodes, due to their distributed topology each requesting node takes longer to find a new node from which to continue downloading the file. This fact causes that file download be slower.

These experiments consider ten malicious nodes to evaluate both infrastructure against a Sybil attack. Malicious nodes are randomly generated, and each of them in turn generates five false nodes in the network. Sybil attack affects the seeder peers in the P2P social network based on BitTorrent and Kademlia. Sybil attack reduces

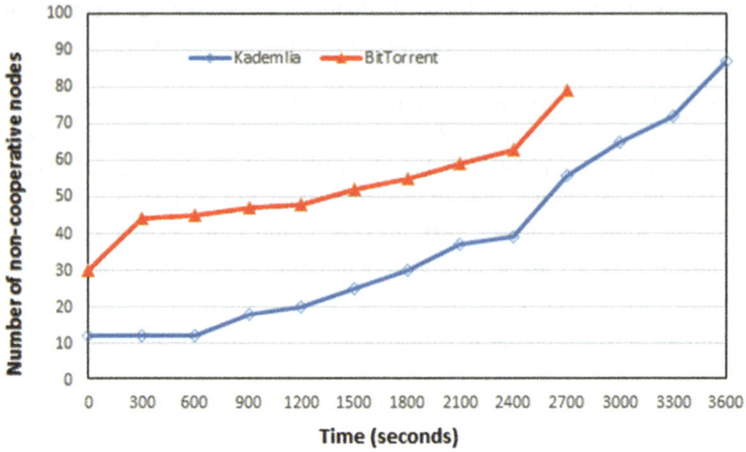

Fig. 3 Performance of P2P infrastructure in the presence of noncooperative peers [36]

the number of seeder peers during a period of time. The decrement of seeder peers has a negative impact on the performance of OSN because it is the seeder peers where the content is obtained. During a time period, leecher peers are increased, because noncooperative peers exploit the vulnerability of the network to obtain benefits from social network such as resources or services without paying for the cost of the benefits. During a Sybil attack the social network may also suffer from other damages such as data alteration or route modification, which are additional problems to the free rider problem.

Figure 4 shows how seeder peers are affected in both networks. Based on these results, we can observe that when the Sybil attack occurs, the BitTorrent network is affected and file begins to download more slowly. Also, download time is very variable at the beginning. However, as malicious nodes go down their reputation, they are eliminated by the tracker and the tracker resources can be released. Thereby, the network can regain continuity and download time is more quick. In this case, the drop nodes are those nodes created by malicious nodes. Figure shows that there are a fewer nodes in the network, which still negatively affects network performance. However, reputation allows that the system can distinguish if a node is uncooperative and remove it from the system. On the other hand, in a social network based on Kademlia, the Sybil attack greatly affects the network both in download time and in the isolating nodes. Many nodes are isolated and others are disconnected from the network; this leads to multiple nodes losing their reputation level because they do not have anything to share, and others leave the network because they have a certain number of requests not answered. If a node has a great number of requests not answered, this node leaves the network too. In these experiments both P2P networks are less vulnerable using reputation levels, as they detect and eliminate (or isolate) those noncooperative nodes. In this case Kademlia is more vulnerable than BitTorrent during a Sybil attack.

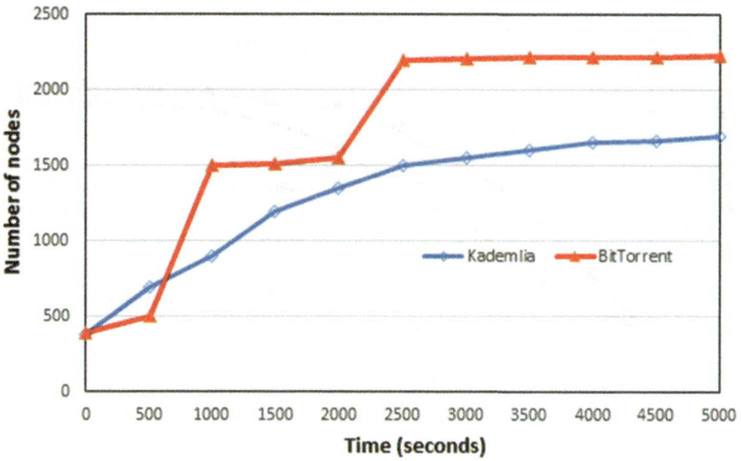

Fig. 4 Comparison of the reduction of the seeder peers in both P2P infrastructure during a Sybil attack [36]

A spoofing attack evaluation is presented in [37] for Kademlia and BitTorrent schemes. In Fig. 5 is compared the performance of Kademlia scheme with and without spoofing attack. Results show that spoofing attack affects the download time. However, there are not many affected nodes, because only those nodes are affected that have been generated with the same identifier (ID), which is what defines them as authentic nodes within the network. That is why the system slows down a bit, because these nodes do not want to share files, and they only want to be downloaded. These facts make the system to reduce the reputation levels of these nodes, so that these nodes are ignored or isolated from the network. The test for BitTorrent scheme under spoofing attacks shows that when there are two or more nodes with the same identifier (ID), the tracker takes an infinite loop and stops managing resources across the network. In other words, the tracker stops working and the network goes down completely because the tracker fails to send a list of neighbors to the nodes that the tracker has connected, as this has stopped working and so the nodes have no neighbors with whom they can communicate.

File integrity is very important in structured and semi-structured P2P schemes, because if a file is modified by any node the tracker loses reference of this file because this reference is maintained via a hash table whose key has been modified too.

6 Conclusions

Decentralized OSN based on P2P networks offers different benefits such as decentralized control, ad hoc deployment, or difficulty taking down. This chapter reviews decentralized architectures for OSNs, and their performance in the presence

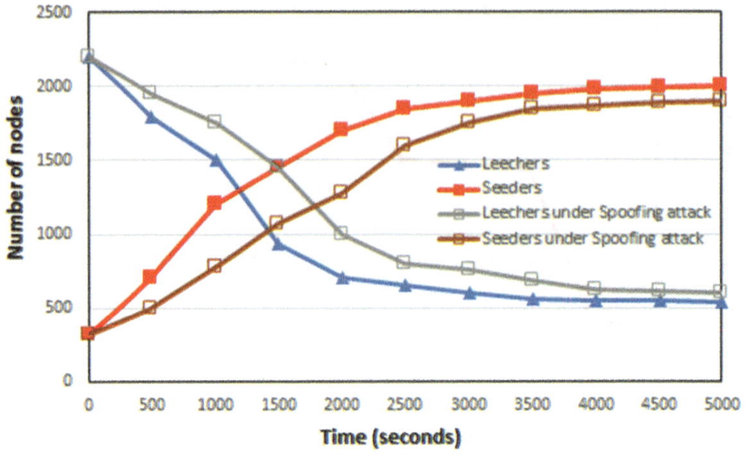

Fig. 5 Performance comparison for Kademlia with and without spoofing attack [37]

of security attacks such as Sybil and spoofing attacks. Specifically, an OSN called Tribler is reviewed for two different P2P infrastructure: hybrid (superpeer) and distributed. BitTorrent protocol was used to simulate the superpeer infrastructure, while the Kademlia protocol simulated the distributed infrastructure. Both ONS schemes were evaluated in terms of delivery time and robustness during a Sybil and spoofing attack. Results show that BitTorrent presents a best delivery time than Kademlia, and it is more vulnerable than BitTorrent against a Sybil attack. However, BitTorrent is more vulnerable than Kademlia against a spoofing attack.

References

1. J.A. Pouwelse, P. Garbacki, J. Wang, A. Bakker, J. Yang, A. Iosup, D.H.J. Epema, M. Reinders, M.R. Van Steen, H. Sips, TRIBLER: A social-based peer-to-peer system, in *Journal Concurrency and Computation: Practice & Experience*. Vol 20(2), pp. 127–138, John Wiley and Sons Ltd, Chichester (2008)
2. S. Buchegger, A. Datta, A case for P2P infrastructure for social networks—opportunities and challenges, in *6th International Conference on Wireless On-Demand Network System*, Snowbird, UT (2009), pp. 161–168
3. M. Durr, M. Maier, F. Dorfmeister, Vegas—a secure and privacy-preserving peer-to-peer online social network, in *4th International Conference on Social Computing* (2012)
4. G. Wondracek, T. Holz, E. Kirda, C. Kruegel, A practical attack to de-anonymize social network users, in *IEEE Symposium on Security and Privacy* (2010), Pp. 223–238
5. B. Krishnamurthy, C.E. Wills, On the leakage of personally identifiable information via online social networks, in *SIGCOMM Computer Communication Review*, vol. 40 (2010), pp. 112–117
6. A. Datta, S. Buchegger, L.H. Vu, T. Strufe, K. Rzadca, Decentralized online social networks, in *Handbook of Social Network Technologies and Applications*, Springer (2010), Pp. 349–378
7. D. Milojicic, D. Valogeraki, R. Lukose, K. Nagaraja, J. Pruyne, B. Richard, S. Rollins, Z. Xu, Peer-to-peer computing, in *Technical Report HPL-2002-57(R.1). HP Laboratories*, Palo Alto, CA (2003)

8. S.R. Chowdhury, A.R. Roy, M. Shaikh, K. Daudjee, A taxonomy of decentralized online social networks, in *Peer-To-Peer Networking and Applications* (2014), Pp. 1–17
9. T. Paul, A. Famurali, T. Strufe, A survey on decentralized online social networks, in *Computer Networks*, vol. 75, pp. 437–452 (2014)
10. K. Goutsias, G. Spathoulas, I. Anagnostopoulos, Testing scalability and performance of decentralized online social networks, in *20th Pan-Hellenic Conference on Informatics*, Patras (2016)
11. M. Castro, M. Costa, A. Rowstron, Peer-to-Peer Overlays: structured, unstructured, or both? in *Technical Report MSR-TR-2004-73, Microsoft Research* (2004)
12. F.A. López-Fuentes, Video Multicast in Peer-to-Peer Networks (Verlag Dr. Hut, Munich, 2009), 138 Pp,
13. S. Kaune, R.C. Rumín, G. Tyson, A. Mauthe, C. Guerrero, R. Steinmetz, Unraveling BitTorrent's file unavailability: Measurements and analysis, in *10th International Conference on Peer-To-Peer Computing*, Delft (2010)
14. D.S. Menasche, A. Rocha, B. Li, D. Towsley, A. Venkataramani, Content availability and bundling in swarming systems, in, *5th International Conference on Emerging Networking Experiments and Technologies*, Rome (2009)
15. J.A. Pouwelse, P. Garbacki, D. H. J. Epema, H. Sips, The Bittorrent P2P filesharing system: Measurements and analysis, in *4th International Workshop on Peer-toPeer Systems*, Ithaca, NY (2005)
16. W. Zheng, X. Liu, S. Shi, J. Hu, H. Dong, Peer-to-peer: A technique perspective, in *Handbook of Theoretical and Algorithmic Aspects of Ad Hoc, Sensor, and Peer-to-Peer Networks*, Auerbach Publications (2006) pp. 587–588
17. P. Maymounkov,D. Mazieres, Kademlia: A Peer-To-Peer Information System Based on the Xor Metric, in *International Workshop on Peer-to-Peer Systems*, Cambridge, MA (2002)
18. Diaspora. http:lljoindiaspora.com/ (2014)
19. A. Bielenberg, L. Helm, A. Gentilucci, D. Stefanescu, H. Zhang, The growth of diaspora—a decentralized online social network in the wild, in *Global Internet Symposium*, pp. 13–18, Orlando, FL (2012)
20. Buchegger, S., Schieberg, D., Yu, L.H., Datta, A., Peerson: P2P social networking - early experiences and insights, in *2nd ACM Workshop on Social Network Systems*, pp. 46–52, Nuremberg (2009)
21. S. Nilizadeh, S. Jahid, P. Mittal, N. Borisov, A. Kapadia, Cachet: A decentralized architecture for privacy preserving social networking with caching, in *The 8th Int. Conference on Emerging Networking Experiments and Technologies* (2012)
22. N. Kourtellis, Finnis, J. Anderson, P. Blackburn, J. Borcea, C. Iamnitchi, A. Prometheus, User-controlled P2P social data management for socially-aware applications, in *11th International Conference on Middleware*, pp. 212–213, Bangalore (2010)
23. Jahid, S., Nilizadeh, S., Mittal, P., Borisov, N., Kapadia, A.,DECENT: A decentralized architecture for enforcing privacy in online social networks, in *4th International Workshop on Security and Social Networking*, Lugano (2012)
24. P. Garbacki, A. Iosup, D.H.J. Epema, M. Van Steen, 2Fast: Collaborative downloads in P2P networks, in *6th IEEE International Conference on Peer-To-Peer Computing*, Cambridge (2006)
25. L. Cutillo, R. Molva, T. Strufe, Safebook: A privacypreserving online social network leveraging on real-life trust. Commun. Mag. IEEE **47**(12), 94–101 (2009)
26. B. Bohen, Incentives build robustness in Bittorrent, in *1st Workshop on Economics of Peer-To-Peer Systems*, Berkeley, CA, (2003)
27. P. Mittal, M. Caesar, N. Borisov, X-Vine: Secure and pseudonymous routing using social networks, in *Network Distributed System Security Symposium*, San Diego, CA (2012)
28. H. Yu, P.B. Gibbons, M. Kaminsky, F. Xiao, SybilLimit: A near-optimal social network defense against Sybil attacks, in *IEEE Symposium on Security and Privacy*, Oakland, CA (2008)
29. B. Viswanath, A. Post, K. P. Gummadi, A. Mislove, An analysis of social network-based sybil defenses, in *SIGCOMM Computer Communications Review*, Vol. 40(4), pp.363–374 (2010)

30. S. Roos, T. Strufe, A contribution to analyzing and enhancing Darknet routing, in *International Conference on Computer Communications (INFOCOM)*, Turin (2013)
31. Y. Chen, W. Trappe,R.P. Martin, Detecting and localizing wireless spoofing attacks, in *4th IEEE SECON* (2007)
32. Y. Sheng, K. Tan, G. Chen, D. Kotz,A. Campbell, Detecting 802.11 MAC layer spoofing using received signal strength, in *IEEE INFOCOM*, Phoenix, AZ (2008)
33. V. Velasco, Introduction to IP Spoofing. SANS Institute InfoSec Reading Room (2000)
34. H. Wang, C. Jin, K. Shin, Defense against spoofed IP traffic using hop-count filtering, in *IEEE/ACM Transactions on Networking*, vol. 15, no. 1 (2007), pp.40–53
35. Z. Duan, X. Yuan, J. Chandrashekar, Controlling IP spoofing through Interdomain packet filters, in *IEEE Transactions on Dependable and Secure Computing*, vol. 5, no. 1 (2007), pp. 22–36
36. F.A. López-Fuentes, S. Balleza-Gallegos, Evaluating Sybil attacks in P2P infrastructures for online social networks, in *IEEE 7th International Symposium. on Cyberspace Safety and Security* (2015)
37. F.A. López-Fuentes, S. Balleza-Gallegos, Investigation of effects of spoofing attacks in P2P online social networks, in *IEEE 30th International Conference on Advanced Information Networking and Applications Workshops (WAINA)* (2016)
38. J.R. Douceur, The Sybil attack, in *1st IPTPS, LNCS*, Springer-Verlag, 2429 (2002)
39. H. Yu, M. Kaminsky, P.B. Gibbons, A. Flaxman, SybilGuard: Defending against Sybil attacks via social networks, in *SIGCOMM'06*, Pisa (2006)

We Shall Not Only Survive to the Future of Social Networks

Christophe Thovex

Abstract Networks of social interactions and personal relationships are now digitized within the social Web. It induces new societal uses and new analytic processes such as viral marketing, but surveillance could also play a major role in the future of social networks.

The story of social networks analysis started in the early 1930s with Moreno's work introducing sociograms and sociometry for the study of group dynamics. It found a renewal in the ability to treat massive data thanks to computers, more than 50 years later.

We propose a double-sided view of the story of social networks, exposing both the economic and societal stakes tied by the common role of specific analytics, toward the multiple futures of social networks.

1 Origins of Social Networks Analysis

We commonly restrict the idea of social networks to well-known parts of the social Web digitizing social networks, but social networks are mostly internal networks hidden to the social Web. When we talk to our neighbors, interact with others for a sport activity, or exchange mails for professional reasons, we are members of internal social networks. Digitizing social networks can foster the development of internal networks, while it enables to collect, connect, and analyze massive and heterogeneous data too—*i.e.*, Big Data.

C. Thovex (✉)
French-Mexican Laboratory on Computer Science and Control (LAFMIA - UMI CNRS 3175), Universidad de Las Americas, Puebla, Mexico, DATA2B, Rennes, France
e-mail: christophe.thovex@data2b.net

T. Özyer et al. (eds.), *Social Networks and Surveillance for Society*, Lecture Notes in Social Networks, https://doi.org/10.1007/978-3-319-78256-0_6

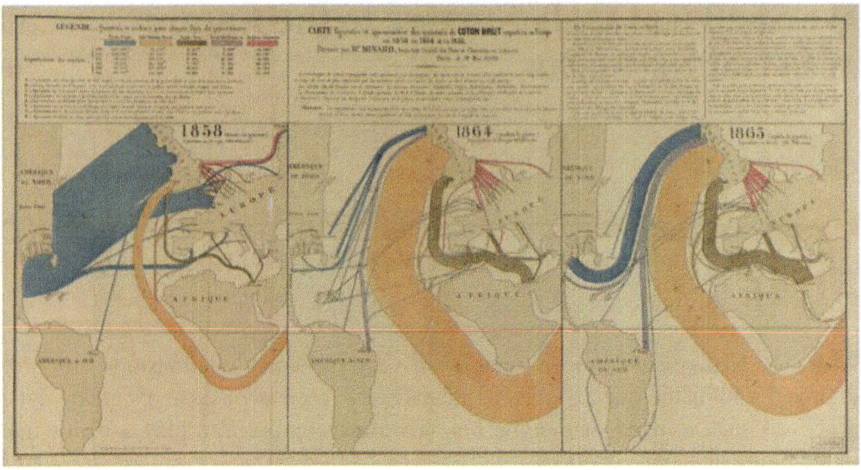

Fig. 1 Influence of the Civil War on the European cotton importations

Sociograms and Sociometry

In 1934, J.L. Moreno introduced the use of sociograms for the analysis of scholar groups interactions and behaviors [18]. Moreno's sociograms were focused on the study of leadership, nexus, and group dynamics. They contributed to social psychology by providing an intuitive way for visualizing the organization of social roles and positions within collectives. They are commonly accepted as an origin of social networks analysis, but graph notions were yet below data representations figuring out socioeconomic interactions, such as in Fig. 1 for the European importations of cotton and their origins during the Civil War (USA)[1].

The graph-based drawing concepts yet used in 1866—Fig. 1—are now implemented in software libraries such as D3JS.[2] It enables the Web visualization of contemporary socioeconomic interactions. For instance, Uber rides during 2012 in the neighborhoods of San Francisco[3] are depicted in Fig. 2.

Social Psychology and Econometry

According to R. Brown, social psychology studies the behaviors and mental processes determined by interactions with other individuals [5]. At the early

[1]Source: http://scimaps.org/mapdetail/europe_raw_cotton_im_3

[2]https://d3js.org/

[3]https://bost.ocks.org/mike/uberdata/

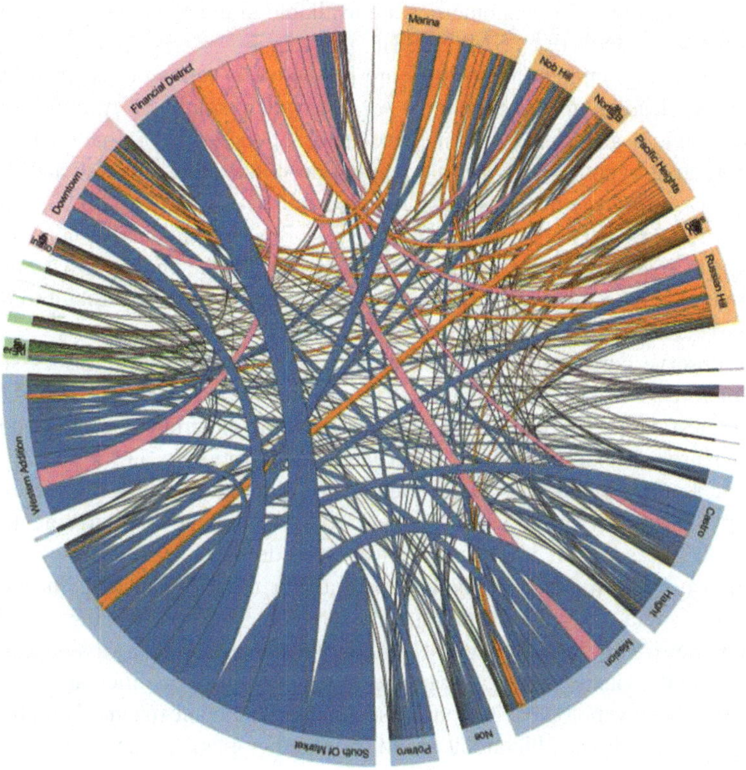

Fig. 2 Uber rides in San Francisco (2012)

beginning of social psychology, G.W. ALLPORT included imaginary and implicit interactions in the definition of social psychology [1]. It is important enough to be underlined, because digital processes do not capture data related to imaginary interactions and as a result, digital analysis remains limited (by explicit tracks) to the inference of implicit interactions, when facing issues in social psychology. However, sociograms and the digital representations of large social networks in graph structures bring new openings in social psychology and sociometry.

Econometrics is a quantitative branch of the economic science based on the estimation and experimentation of economic models thanks to computer science, statistics, and mathematical approaches. It is mostly based on the reuse of methods defined in descriptive statistics and inferential statistics, such as linear regression for the estimation and prediction of values between correlated variables. Tests of independence and significance such as chi-squared and Fisher's test, time-series and classifiers such as support vector machine (SVM) and principal components analysis are subjects to numerous enhancements in the specialized literature. For instance, in [27] the authors present a logistic regression model outperforming the previous models by reducing the error dispersion with multivariate predictors.

Numerous packages for econometrics are available in the open-source platform R, as in proprietary tools (*e.g.*, SAS, SPSS).

Unfortunately, inferential statistics are restrictive in terms of accepted hypothesis, easily leading to false beliefs. As a trivial example of *type II error*, in popular random games with infinitesimal odds of success (*e.g.*, national lottery), the alternative hypothesis is "people playing do not win the jackpot" and the null hypothesis "players win the jackpot" is rejected with $pvalue \leq 0.05$, although there are regularly jackpot winners.[4] Conversely, Bayesian models are not restrictive in terms of hypothesis acceptance, but still depend on the choice of assumptions, possibly leading to false beliefs too. Therefore, it is particularly difficult to rely on standard econometrics when dealing with Big Data.

We observe current trends in machine learning tackling these issues such as random forests and decision graphs, formally closer to neural networks than to inferential statistics [4]. In recent works, we also have experimented a hybrid model merging probabilistic graphs with the Markovian paradigm and a convolutional heuristic for providing subsets of conjunctive assumptions in terms of explaining/explained variables and values (private work—confidentiality agreement). Such an approach was already identified as a hybrid model in [26].

Our economic models are both a part and a result of societal interactions. Therefore, social psychology might be a precious key for opening new tracks in econometrics. J. VON NEUMANN implicitly advocated such a paradigm when introducing the Game Theory, in which decision making is qualified and quantified depending on psychological reasons in socioeconomic contexts, and is represented by extensive forms (*i.e.*, hierarchical and weighted graphs) [10, 25].

Digital Networks and Social Networks Analysis

From the Byzantine Empire to the Second World War, information mastering has played a major role in the evolution of humanity. Turing's work was decisive for exiting the war in 1945. Although they are now considered of public utility, the Internet and the TCP/IP protocol were first a weapon system, then it turned into the tremendous information system that we know, fostering knowledge mapping and sharing thanks to search engines indexing the worldwide content. Nowadays it also digitizes social content and interactions within social networks platforms.

Networks are omnipresent in the dimensions of the World Wide Web. The Web structure is a graph and Google Page Rank is a graph-based algorithm. According to A. SHINGAL (Head of retrieval at Google until 2016), the text of web pages is processed by information retrieval algorithms stemming from models defined in [19, 20] and [21]. Google knowledge graph appears as a semantic network depending on the unstructured content of the Web, allowing the Google engine to

[4]And "100% of losers are players."

process named entities as things and not as strings. Since 2015, Google reinforced its works in machine learning and deep learning. It has released TensorFlow for conversational search[5] and RankBrain, described as the third component of Google engine. Conversational computing is a major trend in Artificial Intelligence (AI). As the Web content, Google knowledge graph and the convolutional neural networks implemented in deep learning are graph-based data structures [16].

Social networks are also graph-based data structures. With the development of computer science since 1980, Social Networks Analysis (SNA) has found a renewal due to the increasing ability to process larger networks in shorter times [8]. Centrality measures and graph-clustering algorithms are now available in open-source libraries as in graph visualization software such as Tulip and Gephi[6] [2]. The oldest centrality measures were based on homogeneous graphs, mostly undirected social graphs such as $G(V, E)$, with V set of vertices representing people and E set of edges representing their social ties [9]. Thanks to the recent interest in the social Web, these algorithms are continuously enhanced and we found now semantic centralities as epistemic models borrowing to multiple domains [3, 23, 24].

Google Knowledge Graph is not just rooted in public sources such as Freebase, Wikipedia, and the CIA World Factbook. It is augmented thanks to information retrieved from the Web by machine-learning algorithms, and tuned based on what people search for. Thus, Google's features are now self-learning from the Web content and from users' interests. It benefits users, and by providing a better access to information, it globally fosters the socioeconomic development. Logically, the future of social networks analysis should also evolve toward self-learning from the Web content and group dynamics, accepting sociological notions such as crowd wisdom and fostering the social cohesion and its corollaries [11, 22].

2 Opportunities and Threats of the Socio-Semantic Web

The semantic Web is based on XML structures such as RDF, SKOS, OWL, SWRL, or FOAF. It is implemented in major components of the Web such as DBPedia, the NYT articles database, and Web search engines architecture. The Web and the social Web are distinct networks of information which are partially hermetic one another. The social Web content needs access controls, thus it is a part of the deep Web, not accessed from the Web search engines. However, the public parts of the social Web are indexed and private parts include back-links toward web pages. Therefore, they self-complete and as their expansion in terms of volumes and technicality benefits from each other, we can talk of the *socio-semantic Web* as a paradigm merging Big Data and AI.

[5] Wolfram Research also develops a computational knowledge engine.

[6] http://www.kdnuggets.com/2015/06/top-30-social-network-analysis-visualization-tools.html.

Such a paradigm is a major stake when viral marketing and emotional analysis gently replace econometrics for decision-making. It means decision support systems are no more limited to marketing applications, but they can process social issues and political issues related to digital populations. It also reminds us the growing effects of digital divide on the main part of the world populations compared to extremely digitized countries mastering digital knowledge, Big Data technologies, and Data Science. As a corollary, silent weapons such as information networks and cryptography—*cf.* Turing's Machine—are augmented with analytic intelligence and essential in governments strategies, from now on.

Crossed Communication for Global Development

Thanks to high-speed networks infrastructures, even mid-class homes in emergent countries are permanently connected. The Y generation listens to playlists of video clips streamed from YouTube, defined from a smartphone, and converging toward a wide connected TV. We share favorite bookmarks with our friends while viewing content, and all our interactions with connected devices are tracked, analyzed, and monetized, sustaining the development of new features in the knowledge economy.

Soon, big data flows will be collected from the connected sensors and objects of the Internet of Things (IoT). Wearable chips will characterize our acts, our emotions, our health, our *quantified self* as a big picture of our life conditions. These crossed flows will contribute to global development. As they belong to Internet companies and telecommunications operators, we might also expect an evolution towards a *monetized self*, coexisting with digital free services. Nevertheless, the full access to Big Data is also full access to economic and political power, entailing a paradigm of total surveillance in which social and semantic networks are key structures for reasoning and analytic intelligence.

Total Surveillance in Digital Humanities

In [7], the philosopher M. FOUCAULT presents panopticism as an enterprise of culpability normalization for individual consciences, through an insight almighty. The panopticon is an economical jailhouse radically changing the power paradigm. With punishment, the smallest number controlled the larger one being visible. With panopticism, the smallest number controls the larger one staying hidden in a modern ivory tower. Figure 3 presents a picture of the panopticon architecture, as designed by the Bentham brothers in 1780.[7] In such a carceral edifice, the surveillance can be delegated to survey as they never know if someone is present in the tower or not.

[7]Source: www.letemps.ch/opinions/2015/02/12/nouvelles-normes-transparence.

Fig. 3 Panopticon

Surprisingly, that principle is an underlying feature of the terrorist events in Paris and Nice (France, 2015 and 2016) inducing fear-based collective behaviors close to panopticism within urban populations.

Are Facebook users enjoying a fantastic freedom of expression or submitted to a power which incites them to disclose their private life?

With social platforms such as Twitter, we might talk of *digital panopticism*, in which users would be guards and prisoners, watching and ranking one another through social content. However, the social Web is much more than a virtual panopticon when it enables people to find each other around the world, crowds to help each other from a country to another, in case of disasters. This is where we shall not only survive to the future of social networks, as a nod to [17].

The social Web is also a wonderful tool for education. Massive Open Online Courses (MOOCs) are a specific resurgence of social content sharing, and a tremendous progress for accessing knowledge from disadvantaged zones, since an Internet connection is available. Unfortunately, panopticism cannot be forgotten. While disadvantaged zones result sometimes from the domination of the poorest by the richest, MOOCs could nevertheless turn into massive open online surveillance of the future generations in dominated countries. In such a case, the discernment of fundamental intentions from each part is essential while the state of the art—Sect. 1—attests of massive capacities to classify, to rank, and to qualify people and their knowledge so as to push viral information for personalized education, as well as for *manufacturing consent* [6, 12, 13].

Digital humanities are commonly defined as the digital data and tools related to the history and knowledge of human civilizations. As a consequence, MOOCs, digital learning, news, and social content such as published in blogs and social networks are a significant part of digital humanities. Terrorism has recourse to communication networks and social media. It stands on digital humanities for its expansion. As a result, counter-terrorism studies digital humanities too and, paradoxically, increases its efficiency proportionally to the expansion of terrorist propaganda as a part of digital humanities. Such a situation feeds a paradigm of total surveillance, through social content and new trends such as the IoT, in the name of opposite beliefs and conflicts.

As a corollary of big data in social and semantic networks analysis, future trends in counter-terrorism could renew some old top-down strategies such as *to search and destroy the hubs first*, with bottom-up ones—*i.e. to cut leaves kills the tree*—in order to annihilate inextinguishable organizations such as ISIS and other nebular systems of corruption, for instance [15].

Individual Information for Collective Empowerment

Whether it be in digital learning or counter-terrorism, Big Data and Data Science for the social and semantic Web make appear ability of fine-grained analysis of massive content [14]. Hybrid graph models for detailed exploration, analysis, and visualization enable to classify, to rank, to detect communities, and to locate individuals within these communities and among millions of users. This growing ability to distinguish each one within a crowd is not necessarily a societal threat. It is also the first opportunity for populations to constitute millions of interests communities aggregating each others and creating new streams of thoughts for which the last remaining frontier is the language barrier. Initiatives for social welfare, petitions, emergency hubs, and groups for daily uses are spontaneously created and adopted by thousands of people every day—e.g., Facebook events, Ecuador earthquake.

Individual information distributed over the world day after day provides a new power to collective actions. If the Wisdom of Crowds presented in [22] is publicly expressed by the *vox populi* everywhere in the world, as long as people will be in measure to follow it and to organize related actions thanks to the social Web, there should be social benefits, economic benefits, and reduction of inequities, as a corollary.

Fostering ties within populations may produce perverse effects such as viral propaganda or coordination of criminal organizations, too. This is where social networks surveillance for society should master the game, and where the wisdom of crowds appears definitely crucial for running a soft revolution, instead of dealing with hard evolution.

3 Societal Requirements and the Future of Social Networks

Historical facts such as revolutions, wars, or scientific discoveries demonstrate how much human evolution is driven by societal requirements. On the one hand, power structures are defined and represented by social organizations, political, financial, philosophical groups owning leadership on populations. On the other hand, populations behold the power on their leading organizations, through votes, public opinions, solidarity, and collective actions. Our hierarchical conception of society is submitted to the decisions of the higher parts of the hierarchy, while these parts fall down since they are not supported by the base majority.

Therefore, the world organization might be roughly represented as a set of hierarchical networks interconnected by a fuzzy subset of common nodes, belonging to one or more of these networks. These multiple hierarchies map various dimensions of societal organizations whether it be governments, global trade networks, transport networks, terrorist networks, or the World Wide Web. Human being is the common denominator to these structures, which as a consequence may appear as a *vastly multidimensional and heterogeneous hidden social network*. Developing semantic and social networks analysis within such a multidimensional network should enable its inner power hierarchies to better listen to the *vox populi* and to understand the wisdom of crowds. Such a paradigm might be supported by global and local recommendation systems based on *deep social learning* as an evolution of social networks analysis and mining. The future of social networks analysis and surveillance is explored in next sections.

Multiple Hierarchies Listening to Wisdom of Crowds

In the thirteenth century, Frederic II of Hoenstaufen left the education of infants to mute nurses, in order to discover the children's natural and innate language. As a result, they never talked and died in the first years of their lives. This historical experimentation demonstrates the absolute necessity of communication for the development of human being and humanity. Learning requires communication, knowledge transmission from generation to generation too, so as the development of cognitive functions seems to depend on communication and social interactions.

Mimetic approaches aim at producing new paradigms and models based on the observation of natural phenomenons and processes. It has already inspired deep learning methods and knowledge networks, for instance. Therefore, we shall mind the openings brought by the newcomers of the digital continuum for future *societal learning* too. *Societal learning* refers here to an extension of deep social learning as a specialization of deep learning and Big Data. The social and semantic Web, Big Data, Data Science, social networks analysis and mining as a part of Network Science may also be considered as digital imitations of some foundations of human cognition. If so, we might expect the hierarchical and multiple dimensions of societal organization to be softly irrigated by analytical flows properly *digitizing the wisdom of crowds*. Our power organization being distributed, and interactive, in-between the tops and the bases of multiple hierarchies, it might result in a new societal paradigm propitious to balance inner dissensions and oppositions and tending to distribute the control within hierarchical structures aimed at more widely shared goals. Such a paradigm could pave the way for a future power regulation to be compared with self-organized anarchy, yet advocated by social trends and beliefs during the years 1970. The term anarchy refers here to its original definition, a political philosophy advocating leaderless societal organization.

Prospective works rarely provide logical proofs but explore the field of possibilities so as to converge into probable hypothesis. The societal paradigm developed hereby, in which multiple power hierarchies listen to the wisdom of crowds, rests on the evolution of trends appearing since 2010 (*i.e.*, Big Data, Data Science, Network Science). The notions of social network and surveillance are inseparable from such a paradigm, in which they definitely appear as significant leverage for the development of society.

Toward Deep Social Learning as a Governance Paradigm

Reminding the history of human civilizations and societies, we retrieve cycles of punctual revolutions succeeding to power representations rising up to dictatorial organizations. The phenomenon is reproduced around the world in various times and places, where power representations are not necessarily human but social ones, such as religious dogmas, philosophical or pseudo-philosophical theories, and economic beliefs alienating populations. Inquisitions, Hitlerian doctrine, and the sumptuous way of life of the latest French Kings collecting the first national tax can be quoted as causes of wars and revolutions, for instance. Historians count 19 revolutions in China since the last three millenniums, mostly due to social imbalances and despotic regimes. These are evidences that humanities have never overtaken this monotonous and stable movement since now.

Digital revolution, social networks, and societal learning enable to expect future changes in societal organization, due to organic and systemic causes similar to those observed in human cognition development since the early years of human being—Sect. 3. If so, it means the current cycle in-between human dictatorial regimes and human revolutions might be open by societal learning, and so this regime might be discontinued.

We state societal learning as a specialization of Artificial Intelligence (AI). In the current regime, two social powers are generally opposed (top *vs.* base of hierarchical organizations), generating that cyclic movement in-between punctual power/establishment rising to dictatorial organization and punctual revolutions/wars. It produces a monotonous and alternative regime in which human strength is opposed to human strength.

In case AI opens the cycle and makes the established regime discontinued, society enters a new regime in which human strength is then opposed to AI strength. Regarding the well-known Moore's law, AI strength likely grows much faster than human strength. Therefore, we might enter a regime in which punctual human revolution/war is quickly dominated by AI strength when it raises to dictatorial organization. Such a regime converges into a fix point in which AI supremacy eliminates the possibility of human revolution/war, entailing a continuous situation of AI dictatorship. Obviously, such a situation does neither fit with the future

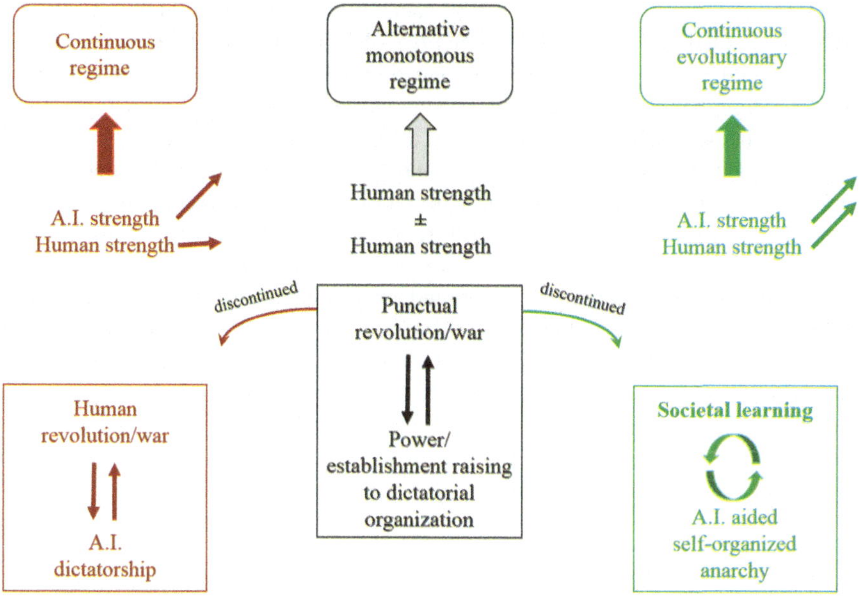

Fig. 4 Societal learning and digital revolution

expressed by the *vox populi* nor with what is expected to find in the wisdom of crowds.

Therefore, societal learning leads AI toward a new and natural regime intrinsically avoiding AI dictatorship in which AI strength pulls up human strength by convolution. Such a regime shall produce a new monotonous and continuous movement of evolution in an AI-aided and self-organized anarchy. Figure 4 illustrates the presented hypothesis.

4 Perspectives, Expectations, and Conclusion

Based on facts and logical considerations presented in the previous sections, we made appear a paradigm of societal learning as a digital and mimetic artifact, reproducing the natural development of human cognition at the societal scale thanks to communication and socialization. It entails social networks as a part of the social and semantic Web, and deep social learning as a part of analytic/artificial intelligence. As a counterpart, surveillance is indissociable of this paradigm.

Society might expect various perspectives tied to digital humanities from the prospective work advocated. Such a work overtakes the interrogations presented in usual pseudo-philosophical studies focused on the future of AI and society. It pushes the prospective until it converges into a deterministic abstract system for the estimation of future societal evolutions caused by digital revolution. As a result, it produces the advocated paradigm in which society enters a new and continuous regime of evolution pushed forward by societal learning, accordingly with trends observed in the human being development, evolution, and expectations.

The digitalization of social ties incarnated by social networks is to be compared with the digitalization of human knowledge within the Web. These memorizations of human being and humanities by digital artifacts, as much simple premises of the future steps of a nascent society they could be, might be seen and understood as technical features as well as societal expectations. However, such an insight would leave us in expectancy regarding the future of AI, social networks, and surveillance in society. Observing causality in-between social ties and the development of intelligence along the human life, we found a deterministic and positive perspective reducing uncertainty about the future of society, tied to digital humanities.

Lastly, appreciating the wonderful ability to develop poetry that characterizes natural intelligence, a citation commonly attributed to *Gandhi* might sagely conclude this chapter reminding us that *It is the law of love which rules humanity. If violence, that is to say hate had governed it, it would have disappeared since a long time*.

References

1. G.W. Allport, The historical background of modern social psychology, in *Handbook of Social Psychology*, ed. by G. Lindzey, vol. 1 (Addison-Wesley, Cambridge, 1954), pp. 3–56
2. V.D. Blondel, J. Guillaume, R. Lambiotte, E. Lefebvre, Fast unfolding of communities in large networks. J. Stat. Mech. Theory Exp. 10008 (2008). http://dx.doi.org/10.1088/1742-5468/2008/10/P10008
3. U. Brandes, D. Fleischer, Centrality measures based on current flow, in *22nd Symposium on Theoretical Aspects of Computer Science (STACS 05)*. Lecture Notes on Computer Science, vol. 3404 (Springer, Berlin, 2005), pp. 533–544
4. L. Breiman, Random forests. Mach. Learn. **45**(1), 5–32 (2001)
5. R. Brown, *Social Psychology* (Free Press, New York, 1965)
6. N. Chomsky, Z. Herman, *Manufacturing Consent: The Political Economy of the Mass Media* (Pantheon Books, New York, 1988)
7. M. Foucault, *Discipline and Punish: The Birth of the Prison* (Pantheon Books, New York, 1977)
8. L. Freeman, Centrality in social networks conceptual clarification. Soc. Netw. **1**(3), 215–239 (1979). http://dx.doi.org/10.1016/0378-8733(78)90021-7
9. L. Freeman, W. Bloomberg, S. Koff, M. Sunshine, T. Fararo, *Local Community Leadership* (Syracuse, New York, 1960)
10. D. Fudenberg, J. Tirole, Extensive form games, in *Game Theory* (MIT Press, Cambridge, 1991), pp. 67–106
11. F. Galton, Vox populi. Nature **75**, 450–451 (1907)
12. O. Gil-Or, Building consumer demand by using viral marketing tactics within an online social network. Adv. Manag. **3**(7), 7–14 (2010)
13. E. Holoan, M. Melia, D. MacMullen, C. Pahl, Adaptive e-learning content generation based on semantic web technology, in *International Workshop on Applications of Semantic Web Technologies for E-Learning* (2005)

14. R.F. Kizilcec, C. Piech, E. Schneider, Deconstructing disengagement: analyzing learner subpopulations in massive open online courses, in *Proceedings of the Third International Conference on Learning Analytics and Knowledge*, LAK '13 (ACM, New York, NY, 2013), pp. 170–179
15. D. Knoke, *Emerging Trends in Social Network Analysis of Terrorism and Counterterrorism* (Wiley, New York, 2015)
16. A. Krizhevsky, I. Sutskever, G. Hinton, Imagenet classification with deep convolutional neural networks, in *Advances in Neural Information Processing Systems 25: 26th Annual Conference on Neural Information Processing Systems 2012*. Lake Tahoe, Nevada (2012), pp. 1106–1114
17. J. Moreno, Who shall survive ? - (Trad. fr) Fondements de la sociométrie. Presses Universitaires de France (1934)
18. J.L. Moreno, *Who Shall Survive: A New Approach to the Problem of Human Interrelations* (Nervous and Mental Disease Publishing Co, Washington, 1934)
19. S.E. Robertson, K. Sparck Jones, Relevance weighting of search terms. J. Am. Soc. Inf. Sci. **27**(3), 129–146 (1976)
20. G. Salton, M. MacGill, Introduction to modern information retrieval, in *Retrieval Refinements*, chap. 6 (McGraw-Hill Book, New York, 1986), pp. 201–215
21. K. Sparck Jones, S. Walker, S.E. Robertson, A probabilistic model of information retrieval: development and comparative experiments. Inf. Process. Manag. Int. J. **36**, 779–808 (2000)
22. J. Surowiecki, *The Wisdom of Crowds* (Anchor Books, New York, 2005)
23. C. Thovex, F. Trichet, Skills networks and professional social networks analysis: from social networks analysis to predictive knowledge analysis, in *Exploratory Analysis for Dynamic Social Networks* (iConcept Press Ltd, Kowloon, 2012), pp. 31–58
24. C. Thovex, F. Trichet, An epistemic equivalence for predictive social networks Analysis, in *Web Information Systems Engineering - WISE 2011 and 2012 Workshops*, Sydney, Australia and Paphos, Cyprus. Lecture Notes in Computer Sciences, vol. 7652 (Springer, Berlin, 2013), pp. 201–214
25. J. Von Neumann, *Theory of Games and Economic Behavior* (Princeton University Press, Princeton, 1944)
26. P. Wilinski, B. Solaiman, A. Hillion, W. Czarnecki, Toward the border between neural and Markovian paradigms. IEEE Trans. Syst. Man Cybern. Part B **28**(2), 146–159 (1998)
27. H. Zou, T. Hastie, Regularization and variable selection via the elastic net. J. R. Stat. Soc. B **67**, 301–320 (2005)

A Centrality Measure for Directed Networks: m-Ranking Method

Reji Kumar and Shibu Manuel

Abstract Understanding the importance of actors (or nodes) in complex networks is an area in social network research. Importance of nodes is interpreted in different ways in different context. If the focus of research is on the spreading of information, then nodes which can spread information throughout the network in a faster pace than any other nodes is important. In the study of spread of disease or virus in a network, the nodes which are at minimum distance from all other nodes are most important. Depending upon the number of direct and indirect connections that each node has in a network, its participation in the spreading varies. Indirect connections specify the nodes that are at two steps, three steps, etc., away from a node. Research for potential spreaders in undirected networks is almost saturated. But in the study of spreading of information, directed networks are more important than undirected networks. In section two of this paper we review some widely used methods to identify important nodes in an undirected network. We also discuss methods which produce similar results for directed networks as well. Wherever necessary we make comparisons of the methods and point out advantages of one method over the other. In section three we propose a method that helps to rank the nodes in a directed network, which takes into account the relative importance of all nodes and directed edges in it.

1 Introduction

Information spreads in a society in two ways. One is the mass media such as news papers, television, etc. Mass media can transfer a significant amount of information direct to a large community. Second is transfer of information from person to person. In this paper we study the transfer of the second type. This method assumes pre-

---

R. Kumar (✉)
P. G. Department of Mathematics, N. S. S. College, Cherthala, Kerala, India

S. Manuel
Department of Mathematics, St. Dominic's College, Kanjirapally, Kerala, India

T. Özyer et al. (eds.), *Social Networks and Surveillance for Society*, Lecture Notes in Social Networks, https://doi.org/10.1007/978-3-319-78256-0_7

existence of some form of relationship between the two individuals who take part in the transfer process. Existence of such a relationship enables us to consider a network of relationship. Thus it is quite natural to assume that the spread of information is significantly affected by the structure of the network and the nature of the members.

Networks that model information diffusion is dynamic. It means structure and other related properties change over time. Changes in relationships and structure are caused by the ever changing nature of human behavior. Dynamic network can be analyzed by comparison of network structures and related properties at different points of time.

With regard to information propagation, the members of a network can be divided into various classes. Most important class contains potential spreaders. Members of this class play a key role in the spreading of information. At the other end of the spectrum of classification there are members who are very dormant or inert in the process. They do not take part or keep silent in the spreading process. All the remaining members come in some class, which lie between the two extremities. Importance of few potential spreaders has been recognized in very large populations [7].

Knowledge of which member belongs to which class is an important factor due to a number of reasons that are context specific. Social scientists wish to control the whole procedure of spreading process in a network for some reasons. Rumors may carry information that is harmful to others. It spreads fast in the communities which have ill will to those who may be affected by it. In the context of spread of disease [1, 12] or virus in a network [2, 8], breaking the path of spread is very crucial in the control procedure. But in some context, we want to increase the pace of spread, because of the benefit the society can get. Public awareness campaigning, election campaigning, advertisements of products are examples.

Thus identifying spreaders according to their ability and strength in the spreading process and various pathways of spreading information have become unavoidable in the control of information transmission in a network. It has led to the discovery of a number of centrality measures and methods to identify potential spreaders. Some important centrality measures are degree centrality, closeness centrality, betweenness centrality, etc. [5]. Degree centrality is the simplest of all centrality measures. It counts the degrees of each node to rank them. Nodes having the highest degree are most important and hence ranked first. Closeness centrality takes into account the length of the shortest paths among nodes in a network. Betweenness centrality quantifies, for each node in a network, the number of times each node appeared as a bridge along the shortest path between any two other nodes. Each of these methods is proven better than other measures in some contexts.

Methods that are recently developed for ranking the nodes in a network are k-core decomposition [6], mixed degree decomposition [21], improved k-core decomposition [10], modified k-shell decomposition [11], etc. In k-core decomposition we find a numerical value for each node. Based on this value the nodes are categorized. Mixed degree decomposition is an improvement of k-shell decomposition method. In k-shell decomposition method we proceed step by step from the nodes having the lowest degree to the next level, after leaving them once its value is calculated. This results in significant loss of information. But in mixed degree decomposition, the number of edges connected to all the removed nodes is also considered. In improved k-shell decomposition method calculations are based on the distance from a target node to the network core. This helps us to distinguish the spreading influence of the nodes within the same k-shell. For weighted networks a method known as weighted k-shell decomposition was suggested by Garas et al. [6]. This method is improved by Kumar et al. in [14]. Generalization of node centrality in weighted networks is the main theme of the paper by Opshal et al. [13]. Very recently Wang et al. developed a method for fast ranking of influential nodes [17] in complex networks. His method improves the k-shell decomposition method using iteration factor. So it is named decomposition using k-shell iteration factor.

Directed network is a network in which direction is associated with each edge. Directed networks are represented by directed graphs. To know more about directed networks, readers may refer [3, 19]. The direction of an edge represents the direction of flow of information from one node to another. In the study of information diffusion, directed networks are very important, as almost all networks of this kind are directed. In many situations transfer of information is one way. People are of various interest. Their likes and dislikes affect the spread of information from one person to another to a great extent. So extensions of the existing methods are essential to handle such situations. Weighted directed graphs are suitable extensions of ordinary graphs, which can handle this situation effectively. Weights associated with each edge in a directed network signifies either intensity, speed or strength of flow of information.

Centrality of directed networks is the focus of attention of some social scientists in recent years. White et al. [20] generalized Freemanan's [4] geodesic centrality measures for betweenness on undirected graphs to the general context of directed graphs. In a recent paper Sims et al. [16] investigated importance of critical nodes or middle men in social or economic networks with reference to the flow of information or trade. In this paper the concept of critical nodes is extended to directed networks. Strong and weak middle men are identified in the network. It is based on node contestability which is introduced as a form of competition in networks.

In the following section we will go through some important centrality measures of ordinary network, which can be easily extended to directed networks. In section three we will also discuss an extension of m-ranking method, which was proved to be an effective method for ranking all nodes in an undirected network. In section four we summarize the findings.

2 Centrality Measures and Directed Networks

Directed networks differ from ordinary networks because of the directions associated with each edge in it. The direction signifies one way communication, one way traffic of information, unidirectional transfer of commodities, etc. A piece of information that flows in one direction may not flow in the opposite direction. The pace and nature of flow depends on three important factors. One is the nature of the transmitting node and receiving node and another is the nature of the communication channels. The third factor is the nature and content of information. In the current study we concentrate only on the first two aspects.

Directions associated with the edges result into two types of degree in a directed network. These degrees are in degree and out degree. In degree is the number of edges which are directed towards a node and out degree is the number of edges which are directed towards other nodes from a given node. In the transmission of information second type of directed edges are significant than the first type. So we base our study on out degree of nodes in a directed network.

Based on the values of in degree or out degree of nodes, we can draw some conclusions about the nodes. If both degrees are zero, the node is isolated in the network. Such nodes do not play any role in the network. If the in degree is zero and the out degree is greater than one, then the node can transmit information generated by itself only. On the other hand, if the in degree is positive and the out degree is zero, then the node is an accepter of any information. But it does not transmit any information to other nodes. The above-mentioned type of nodes are only partial participants in the process of spread of information. We are more interested in the nodes having both degree non-null.

A piece of information can reach from one node to another if and only if there is some directed path from the first node to the second. All nodes that lie along this path are very important in connection with the communication between the two nodes. If all directed paths from one node to another pass through a particular node in the network, then the node is very critical in the sense that any communication between the nodes is highly influenced or controlled by such nodes. Thus, the role of actors in a directed social network is more or less depended on the in degree and out degree of the nodes.

Existence of two or more node disjoint paths is very crucial in the analysis of the content of information received at a destination. If the information received at a node via different paths differ, correctness of the information is questionable. Most often the transmitted information is mixed with the addition of the intermediary nodes. Thus an information which passes through a cut node in a network is completely under its control.

Next we calculate important centrality measures of a simple directed network (Given in Fig. 1). We then compare the values of centrality measures and hence arrive at a conclusion about the best centrality measure. The network contains 15 nodes and 23 directed edges which describe the direction of flow of information in the network. First we discuss the out degree centrality, which is the counter part of simple degree centrality in which the nodes are ranked in the order of degree of the nodes.

Out degree centrality is based on the number of directed links starting from the node. For the node 1, the out degree is 1. Similarly, out degree of node 2 is 1 and that of node 4 is 2. Rank obtained when the nodes are arranged in the descending order

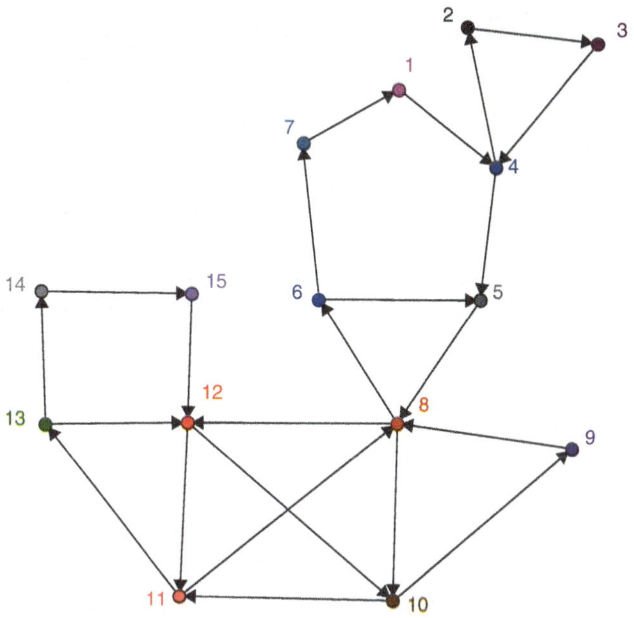

Fig. 1 Example directed network containing 15 nodes

Table 1 Ranking of nodes in Fig. 1 using out degree centrality

Rank	Out degree	Node name
1	3	8
2	2	4,6,10,11,12,13
3	1	1,2,3,5,7,9,14,15

of their out degree is given in Table 1. In the process of spreading information in a large network, the nodes with large out degree will have greater spreading ability. Here, node 8 has the highest out degree and ranked 1 in the spreading ability.

In degree in a directed network has completely different interpretation than out degree. For example, a node having positive in degree and zero out degree can only receive information. He cannot actively take part in the spreading process. Information when passes through many channels may undergo changes in its content or meaning because of the personal biases of the intermediary actors. A node having higher values of in degree is possible to receive an information which originates from one source but travels through many paths, with variations. Such nodes can very well understand these variations. A node having zero in degree and a pretty big value of out degree can be important in a network. But it can transmit only its own information. Thus in degree centrality gives only a polarized view in the spreading process.

We can rank nodes in descending order of their in degree. For the schematic network given in Fig. 1 the nodes are ranked as given in Table 2.

Table 2 Ranking of nodes in Fig. 1 using in degree centrality

Rank	In degree	Node name
1	3	8,12
2	2	4,7,10,11
3	1	1,2,3,5,6,9,13,14,15

The *closeness centrality* measure for undirected graphs was proposed by Freeman [5]. This measure uses sum of distance of all shortest paths from a node to all other nodes. In a similar way in directed graphs, we find sum of distance of all shortest paths from a node to all other nodes of a network and closeness centrality is defined as

$$C_c(i) = \left[\sum_j d(i,j)\right]^{-1}.$$

In the above equation $d(i,j)$ is the distance of the node j from the node i. In the case of a weighted network distance between two nodes i and j is defined as

$$d^w(i,j) = \min\left(\frac{1}{w_{ih}} + \cdots + \frac{1}{w_{hj}}\right)$$

where h are intermediary nodes and w_{ih} denote its weight.

The closeness centrality of a weighted directed graph is defined as

$$C_c(i) = \left[\sum_j d^w(i,j)\right]^{-1}$$

The closeness centrality of the nodes in the directed graph given in Fig. 1 is given in Table 3.

Betweenness centrality of both undirected and directed graphs was also proposed by Freeman [5]. Betweenness centrality of a node in a directed graph is defined as $C_B(i) = \frac{g_{jk}(i)}{g_{jk}}$, where $g_{jk}(i)$ is the number of shortest directed path from node j to node k passing through node i and g_{jk} is the total number of shortest paths from j to k. Betweenness centrality measure of the nodes of the network given in the example is given in Table 4.

Next we discuss another important stratification of the nodes which has recent origin. In 2010, Kitsak et al. put forward a fast node ranking method named *k-shell decomposition* [9] (also called k-core decomposition) for large networks. It partitions the whole node set into subsets of nodes having same rank. Using this method we can find an index k_s for each node, which helps us to know how much important is each node in the network. Nodes with high values of k_s are located in the center or core of the network and nodes with low values of k_s lie in the periphery.

Table 3 Ranking of nodes in Fig. 1 using closeness centrality

Rank	Closeness value of node i	Node i
1	1/41	8
2	1/44	11
3	1/50	4, 6
4	1/52	9,12
5	1/53	7,10
6	1/57	1
7	1/58	13
8	1/62	3,15
9	1/68	5
10	1/72	14
11	1/74	2
12	1/50	6

Table 4 Ranking of nodes in Fig. 1 using betweenness centrality

Rank	Betweenness value of node i	Node i
1	133/238	8
2	88/238	4
3	72/238	7
4	71/238	11
5	69/238	12
6	55/238	6
7	50/238	1
8	47/238	5
9	43/238	13
10	40/238	10
11	22/238	14
12	16/238	3
13	14/238	2
14	13/238	15
15	9/238	9

This way, the network is divided into different layers, which signify the hierarchy of its nodes. This method is extended to weighted networks [18].

The procedure in an ordinary network goes as follows. First we remove recursively from the network all nodes of the network with degree $k = 1$ and we assign the integer value $k_s = 1$ to them. This procedure is repeated iteratively until there are only nodes with degree $k \geq 2$ in the network. Subsequently, we removes all nodes with degree $k = 2$. This procedure is repeated iteratively until there are only nodes with degree $k \geq 3$ left in the network and assign to them the integer value $k_s = 2$ and so on. This procedure is applied until all nodes of the network have been assigned to one of the k-shells. This procedure generates a partition of all nodes of the network into different shells with an integer value. This is how the original k-shell decomposition method works.

The above procedure can be extended to directed network. At each step removal is based on out degree. For the network given in Fig. 1, the k-shell decomposition partitions the nodes as follows. $V_{K_s=1} = \{1, 2, 3, 5, 6, 7, 9, 14, 15\}$, $V_{K_s=2} = \{4\}$, $V_{K_s=3} = \{10, 13\}$, and $V_{K_s=4} = \{8, 11, 12\}$. The set $V_{K_s=i}$ contains all nodes having $K_s = i$.

The k-shell method is a dynamical network decomposition procedure in the sense that the residual degree (number of links connected to the remaining nodes) of nodes is updated in each step while all the information of the removed nodes is dropped. Due to this reason sometimes this method shows serious flaws in ranking procedure. *Mixed Degree Decomposition* (MDD) was proposed by An Zeng et al. in 2013 [21] as a remedy to this problem. In this method both residual degree and exhausted degree (number of links connected to the removed nodes) of the nodes are recorded and the decomposition is based on both of them. In the MDD method, for node i, exhausted degree is denoted by $k_i^{(e)}$ and residual degree is denoted by $k_i^{(r)}$. In each iteration, the nodes are removed according to the mixed degree defined by

$$k_i^{(m)} = k_i^{(r)} + \lambda k_i^{(e)}.$$

In the above expression the symbol λ is a tunable parameter between 0 and 1. Important steps in the decomposition are described below.

1. Initially $k^{(m)}$ for each node is equal to the degree of the node, since there are no removed nodes in the network.
2. Remove all nodes with least degree, denoted by M and assign them to the M-shell.
3. Update $k^{(m)}$ of all the remaining nodes by $k_i^{(m)} = k_i^{(r)} + \lambda k_i^{(e)}$. Then remove all nodes with degree smaller than or equal to M and assign them to the same M-shell. This step is recursively carried on until all remaining nodes are larger than M.
4. Repeat steps 2 and 3 as M value increases until all nodes in the network have been assigned to one of the shells.

When $\lambda = 1$, this MDD method coincides with the degree centrality method and when $\lambda = 0$, this method coincides with the usual k-shell decomposition. The MDD method is no longer integer since $k^{(m)}$ can be decimal when we take λ between 0 and 1. This is the usual MDD method. This can be extended to directed networks by considering out degree of nodes at each time. Partition obtained by applying MDD method to the example network is $V_{K_S=1} = \{1, 2, 3, 5, 6, 7, 9, 14, 15\}$, $V_{K_S=2} = \{4\}$, $V_{K_S=3} = \{10, 13\}$, $V_{K_S=4} = \{11, 12\}$, and $V_{K_S=5} = \{8\}$.

Significance of the MDD method can be understood from the dynamics of virus spread. Suppose virus originates from a node with large exhausted degree. It not only has the same probability as the other nodes in the same shell to infect the nodes in the higher shells, but also it has a bigger branch of nodes in the lower

Table 5 Ranking of nodes in Fig. 1 using improved k-shell decomposition method

K_s value	$\theta(i)$	Node i
1	60	2,5
2	48	1,3
3	36	6,14
4	27	4
5	24	7,9,15
6	12	10,13
7	3	8,11,12

shells to infect, so that this virus will end up covering much more nodes at the end. Thus the information of the exhausted degree cannot be overlooked when ranking the nodes. The k-shell has limited number of ranks and the number of nodes in each rank is quite high, which means that the node differences are not well distinguished in k-shell decomposition. In MDD method, the nodes are more differently ranked than k-shell decomposition.

Further improvements have been made on k-shell decomposition in order to increase its efficiency. In 2013, Liu et al. [10] proposed the improved k-shell decomposition method, in terms of the distance from a target node to the network core. Thus the spreading influence of the nodes within the same k-shell values could be distinguished. The formula for improved k-shell decomposition is $\theta(i/k_s) = (k_s^{\max} - k_s + 1)\sum_{j\in J} d_{ij}$. In the formula, $k_s^{\max}$ is the largest k-core value of the network and d_{ij} is the shortest distance from the node i to the node j. J is the set of nodes whose k-shell value is the maximum. Using this value we can rank the nodes of a network. We can extend this method to directed graphs by considering only the directed path from node i to j. Using this formula in the example network we get the ranking of the nodes as in Table 5.

Degree centrality is the fastest and simplest of all methods. Ranking obtained by this method may not accurately indicate the potential spreaders because degree of the nodes in the neighborhood of a given node does not take part in the calculations. The k-core is also simple to apply. It is an iterative method. It also shares all the disadvantages of degree centrality. Many nodes may get same rank when we use this method. In MDD method exhausted degrees are also considered to rank the nodes of a network. MDD method is better than k-core, because more number of classes is obtained by this method than the number of classes obtained by the other two methods. Improved k-core method and modified k-core method can give much more number of classes than the MDD method. All the above-mentioned methods depend only on the degree of nodes.

In the methods that are already discussed, calculated values of each node center around some neighborhood of that node. But spreading process in a network is more or less influenced by the nature of all nodes and arcs in it. This observation motivates us to define a measure of centrality that has in it the effect of a remote node or edge. This idea is behind the methods proposed by Reji Kumar and Shibu Manuel in 2017 [15] for undirected graphs. This method, which is named *m-ranking*, is very efficient in undirected weighted networks. In this method, for each node the total power is calculated. It is possible to extend this method to weighted directed network, by considering the outward degree and outward wights of nodes. The formula for total power of each node is calculated using the following formula. Since the graph given in Fig. 1 is unweighted, we put $\alpha = 1$ and $\beta = 2$ in the equation given below.

$$T(i) = \left\{ \alpha \left[od_i^{(0)} + \frac{1}{\beta} \sum od_i^{(1)} + \frac{1}{\beta^2} \sum od_i^{(2)} + \cdots \right] \right. \\ \left. + (1-\alpha) \left[\sum oW_i^{(0)} + \frac{1}{\beta} \sum oW_i^{(1)} + \frac{1}{\beta^2} \sum oW_i^{(2)} + \cdots \right] \right\} \quad (1)$$

In the equation α is a parameter between 0 and 1 and $\beta > 1$ is another parameter. Also $od_i^{(0)}$ is the outward degree of the node i and $\sum od_i^{(j)}$ is the sum of the outward degrees of the nodes at a distance j away from node i. Similarly, $\sum oW_i^{(0)}$ is the sum of the weights of the outgoing edges of the node i, and $\sum oW_i^{(j)}$ is the sum of the weights of the edges j distance away from node i. Since we are considering degree of all nodes and weights of all edges, usually the total power of all nodes will be different except for nodes which are distinguishable with respect to isomorphism. When β value is very large this method tends to usual degree centrality. We rank the nodes in the descending order of total power $T(i)$. If we put $\alpha = 1$, this method can be applied to unweighted directed network. Ranks obtained for the example graph is given in Table 6.

Table 6 Ranking of nodes in Fig. 1 using $T(i)$ values

Rank	$T(i)$	Node i
1	7.89	8
2	5.94	12
3	5.89	11
4	5.63	10
5	5.22	13
6	5.09	4
7	5.08	6
8	4.94	7
9	4.82	9
10	4.03	15
11	3.54	1
12	3.41	3
13	2.98	14
14	2.77	5
15	2.58	2

By comparing the ranks obtained by different methods that we have discussed so far it is possible to conclude that the m-ranking method is a very efficient method in ranking the nodes in different classes. Now we proceed to calculate these centrality measures for a real weighted and directed network.

3 m-Ranking for Directed Networks: A Case Study

In the network the nodes represent airports in India and a directed edge from an airport i to j signifies air connectivity of spicejet airlines by at least one flight. Weight of this arc is the number of flights operated by that airlines on a particular day (7 Feb 2017). If at least one flight is operated by the company from the airport j to the airport i, then there is a reciprocating arc. The data for all days are available in the web site of the company www.spicejet.com. Network of the data (Fig. 2) is drawn using Pajek software. The network contains 37 airports (nodes) and 570 routes (directed links) connecting these nodes.

Five different centrality measures of all nodes in the network are calculated using the respective formulae. They are ranked according to the values obtained. The ranks obtained by each node is given in the column under respective name of the centrality against the name of the airport. From the tabulated values it is clear that k-shell decomposition method is the worst method. All other methods can rank the nodes in more or less same fashion. New Delhi is the airport ranked first in all methods. Gorakhpur is ranked last (Table 7).

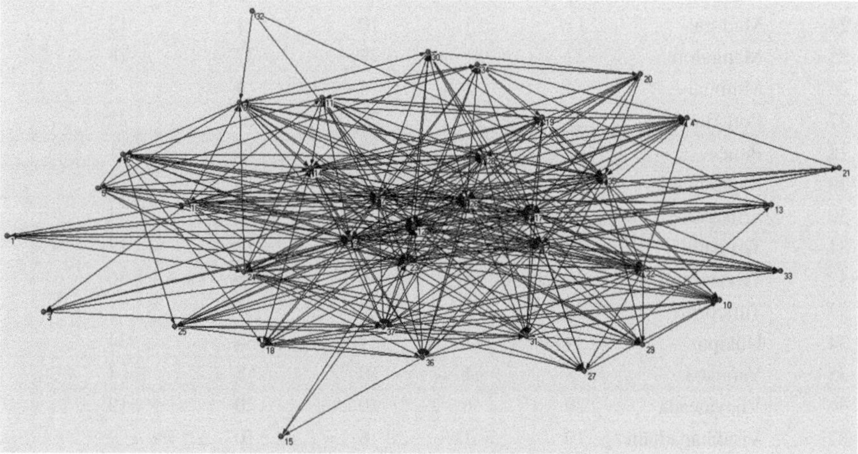

Fig. 2 Directed network of Spicejet flights on 7th February 2017

Table 7 Ranking of nodes using various centrality methods of the network given in Fig. 2

Node label	Airport name	Degree	K-shell	Weighted k-shell	Closeness centrality	m-Ranking
1	Agarthala	24	6	31	24	31
2	Ahemadabad	7	1	7	7	7
3	Aizwal	32	9	33	32	33
4	Amritzar	22	5	26	22	27
5	Bagdogra	33	10	32	33	32
6	Belgaum	30	9	28	30	26
7	Benguluru	2	2	2	2	2
8	Chandigarh	18	7	20	18	30
9	Chennai	5	1	5	5	5
10	Coimbatore	23	6	21	23	20
11	Dehradun	13	2	13	13	13
12	Delhi	1	1	1	1	1
13	Dharmasala	31	9	30	31	29
14	Goa	15	3	15	15	15
15	Gorakhpur	37	14	37	37	37
16	Guahati	9	1	12	9	10
17	Hyderabad	3	1	3	3	3
18	Jabalpur	26	8	24	26	25
19	Jaipur	19	5	17	19	16
20	Jammu	25	7	23	25	21
21	Jodhpur	35	12	35	35	35
22	Kochi	14	1	14	14	14
23	Kolkata	6	1	6	6	6
24	Madurai	11	1	10	11	12
25	Mangalore	27	8	29	27	28
26	Mumbai	4	1	4	4	4
27	Port Blair	21	5	22	21	22
28	Pune	12	3	11	12	11
29	Rajahmundry	16	3	18	16	18
30	Srinagar	28	7	25	28	24
31	Tirupathi	17	4	16	17	17
32	Trivandrum	36	13	36	36	36
33	Tuticorin	29	7	27	29	23
34	Udaipur	34	11	34	34	34
35	Varanasi	8	1	9	8	8
36	Vijayawada	20	6	19	20	19
37	Visakhapatnam	10	1	8	10	9

4 Conclusion and Future Research

Directed networks are very important in the study of dynamics of spreading of information. In a directed network, information can be transferred in a specified direction. Direction in which information passes is highly dependent on the nodes and arcs which lie along the path. Research in this field is yet to be progressed and a lot of work is to be done in the unexplored areas of dynamics of spread of information in weighted directed networks. A directed path along which information can flow may not exist in a directed network, even if such paths exist in the underlying undirected networks. So what we have done for undirected networks are either completely modified or corrected to match with the different situation of directed networks. This challenge increases the enthusiasm and interest of social scientists. In this paper we have initiated a study of identifying potential spreaders or important nodes in a directed network. Some important measures of centrality, which are originally proposed for undirected networks are discussed along with its possible extension in directed networks. An effective ranking procedure named m-ranking for directed network is explained in detail and it is applied to a real network to obtain the ranking of nodes which enable us to identify the important nodes in that network. In the calculations all nodes and edges in a network take part. This is a distinction of this method compared to all other methods. It is also seen that the ranking obtained is perfect. So this method is very suitable for social networks in which speed and quality of the transfer of information varies depending on the nodes and directed edges. In future, all the aforementioned personal biases and variations must be incorporated in this study.

Acknowledgements The first author acknowledges the financial support given by UGC by sanctioning the major research project No. 40-243/2011 SR.

References

1. F. Brauer, P. van den Driessche, J. Wu (eds.), *Mathematical Epidemiology* (Springer, Berlin, 2008)
2. F.B. Cohen, *A Short Course on Computer Virus* (Wiley, New York, 1994)
3. N. Deo, *Graph Theory with Applications to Engineering and Computer Science* (PHI, New Delhi, 1997)
4. L.C. Freeman, A set of measures of centrality based on betweenness. Sociometry **40**, 35–41 (1977)
5. L.C. Freeman, Centrality in social networks conceptual clarification. Soc. Netw. **1**(3), 215–239 (1979)
6. A. Garas, F. Schweitzer, S. Havlin, A k-shell decomposition for weighted networks. New J. Phys. **14** (2012). http://dx.doi.org/10.1088/1367-2630/14/8/083030
7. E. Katz, P.F. Lazarsfield, *Personal Influence: The Part Played by People in the Flow of Mass Communications* (New Brunswick, Transaction Publishers, 1955)
8. J.O. Kephart, G.B. Sorkin, D.M. Chess, S.R. White, Fighting computer viruses. Sci. Am. **277**(5), 56–61 (1997)

9. M. Kitsak, L.K. Gallos, S. Havlin, F. Liljeros, L. Muchnik, H.E. Stanley, H.A. Makse, Identification of influential spreaders in complex networks. Nat. Phys. **6**(11), 888–893 (2010)
10. J. Liu, Z. Ren, Ranking the spreading influence in complex networks. Physica A **392**, 4154–4159 (2013)
11. S. Manuel, K. Reji Kumar, An improved k-shell decomposition for complex networks based on potential edge weights. Int. J. Appl. Math. Sci. **9**(2), 163–168 (2016)
12. M.E.J. Newman, Spread of epidemic disease on networks, Phys. Rev. E **66**(1), 016–128 (2002)
13. T. Opsahl, F. Agneessens, Node centrality in weighted networks: generalizing degree and shortest paths. Soc. Netw. **32**, 245–251 (2010)
14. K. Reji Kumar, S. Manuel, Spreading information in complex networks: a modified method, in *Proceedings of International Conference on Emerging Technological Trends, ICETT 2016*, Bangalore, IEEE Digital Explore Library
15. K. Reji Kumar, S. Manuel, The m-ranking of nodes in complex networks, in *Proceedings of COMSNETS 2017*, Bangalore, IEEE Digital Explore Library
16. O. Sims, R.P. Gilles, Critical nodes in directed networks, 1 Feb 2014. arXiv:1401.0655v2 [cs.SI]
17. Z. Wang, Y. Zhao, Fast ranking influential nodes in complex networks using a k-shell iteration factor. Phys. A **461**, 171–181 (2016)
18. B. Wei, J. Liu, Y. Deng, Weighted k-shell decomposition for complex networks based on potential edge weights. Physica A **420**, 277–285 (2015)
19. D.B. West, *Introduction to Graph Theory* (Prentice Hall, New Delhi, 2006)
20. D.R. White, S.P. Borgatti, Betweenness centrality measures for directed graphs. Soc. Netw. **16**, 335–346 (1994)
21. A. Zeng, C. Zhang, Ranking spreaders by decomposing complex networks. Phys. Lett. A **377**, 1031–1035 (2013)

Searching on Facebook Through the Lens of the Concept of Privacy

Chih Ping Chang

Abstract This chapter points out that judges and legislators didn't consider privacy risks on social networking sites and user's thinking mode in this legal privacy protection mechanism. Social networking sites have considerable privacy risks: improper personal information flow, such as information to be leaked to the third-party software, to be tagged unwillingly on photos, or involuntarily to be shared one's own living news, even self-monitoring by individual recognizes while using social networking sites, or social control. And a user is accustomed to assess privacy risks and set privacy preferences based on his/her experiences and social relationships in the real world. This thinking mode is different form the legal privacy protection mode (on/off), and could not be sufficiently protected by the current regulations. This chapter thought that Helen Nissenbaum's "privacy in context" theory may be the solution. This theory not only considered the approach that users assess privacy risks, but also provided "appropriateness" and "transmission" principles to solve the current issues. In fact, privacy as context theory did not create a totally new theory deviating from the reasonable expectation of privacy. It is better to say that this theory, as the supplementary, gave a new approach to explain the reasonableness standard, and a new normative sense of the past legal concept. It might find a more reasonable boundary to protect the gradually eroded privacy on social networking sites.

1 Introduction

During these days, sharing information on social networking sites is like a routine for many people in their daily life. Information is gathered together in these constantly thriving social networking sites, and also it brings together much helpful information for the criminal investigation or prosecution. Thus, some issues are

C. P. Chang (✉)
Shih Hsin University School of Law, Taipei, Taiwan
e-mail: ccp@mail.shu.edu.tw

T. Özyer et al. (eds.), *Social Networks and Surveillance for Society*, Lecture Notes in Social Networks, https://doi.org/10.1007/978-3-319-78256-0_8

raised: Could the police directly search information a person posted publicly on his/her Facebook? If a person limited his/her Facebook information only to his/her friends, could the police ask cooperation of one's Facebook friend or just pretend a friend to join one's Facebook, in order to search this one's Facebook trying to obtain information for criminal cases? For the government investigators, freely accessing the information on social networking sites is really useful and effective to prosecute crimes. But for the users, comparatively they are not happy to see their information to be used beyond the scope originally envisaged, even out of their control. As the intermediary that privacy guarantees individual freedom of online activities from interests of criminal investigations and prosecution, the current legal system has established a mechanism for keeping a private sphere out from the government's hands. Besides, each social networking site has established its own mechanism for privacy protection, considering information security and user privacy in general way. However, here are some doubts, whether these privacy protection mechanisms successfully prevent arbitrary investigations from invading individual space on websites, or they are artifacts only to provide imaginaries that users can control their information, but in fact the protection of privacy is pierced through and eroded continuingly.

This chapter focuses on issues on criminal search on social networking sites and privacy protection in the context of American legal system, and attempts to point out that utilizing privacy settings on social networking sites, in conjunction with "privacy as context" theory, in fact, can help solve issues of privacy violations. Furthermore this chapter tries to explain an expectation gap among legal system, social networking sites, and users from the perspective of privacy risk. This gap just highlights the difficulty of building the privacy protection mechanism on social networking sites. Therefore, the structure of this chapter is first to depict the privacy protection mode in legal system, then to point out the risk of privacy and protection mechanisms on social networking sites, then to analyze the current situation by cases, and finally to suggest a possible solution.

2 Reasonable Expectation of Privacy and Legal Search

When law enforcement agencies want to obtain Facebook evidence, as the same way when they wants to obtain general evidence, the government should not cross the boundary drawn by the Fourth Amendment, that is, the so-called reasonable expectation of privacy. To protect people's right against the governmental arbitrary investigation, the most important principle guaranteed by the Fourth Amendment of American constitution stated, "The right of the people to be secure in their persons, houses, papers, and effects, against unreasonable searches and seizures, shall not be violated, and no Warrants shall issue, but upon probable cause, supported by oath or affirmation and particularly describing the place to be searched, and the persons or things to be seized."[1] Based on that, the American Supreme Court built a privacy protection mechanism from illegal search and seizure by police in the Katz case.[2] This case discussed the issue on that the police wiretapped the public telephone booth while the defendant, Katz, was communicating to place the illegal wagers, and then used this gathered information to justify Katz's arrest and conviction. The majority argued that "what a person knowingly exposes to the public, even in his own home or office, is not a subject of Fourth Amendment protection."[3] Except this "knowingly exposed to the public" rule, Justice Harland in the concurrence provided a two-pronged test for reasonableness of privacy expectations to decide when to appropriately apply the Fourth Amendment in a governmental investigative search. According to Harland's theory, the range guaranteed under the Fourth Amendment should be decided by the person's reasonable expectation of privacy. This so-called reasonable expectation of privacy asks for a twofold requirement: the first is "a person has exhibited an actual (subjective) expectation of privacy" and, second, "that the expectation be one that society is prepared to recognize as 'reasonable'."[4] This two-pronged test is applied broadly and the court still uses today.[5]

Therefore, a legal search must meet the following test to ensure no government investigation against people's constitutional rights. Governmental investigators must first confirm whether their actions constitute conducts of search under the Fourth Amendment. That is, it must first determine whether the person searched by the police has a reasonable expectation of privacy. If the person has a reasonable expectation of privacy in this case, the police need to obtain a warrant to conduct the search; otherwise, the search is illegal. On the contrary, the search without a warrant will be legal, while this person has no reasonable expectation of privacy. Second, in the case of reasonable expectation of privacy, the investigators conduct a legal search with warrant. This warrant is issued by the judge, requested a probable cause, and described the range of this search or seizure in detail. Third, in the case of reasonable expectation of privacy, search without a warrant is not absolutely illegal. If the case fell into exceptions recognized by precedents, such as consent search, exigent circumstances, plain view doctrine, and search incident to lawful arrests, the investigators can search the person legally without a warrant. Search without a warrant is an exception, and can only be built by judgments. The last situation is in the case of reasonable expectation of privacy where the investigators did not obtain a warrant for search, nor find this situation falling into any exceptions. This search is illegal, and may comply with the exclusionary rule, leading to exclude the obtained evidence in this search.

---

[1] U.S. Const. amend. IV.

[2] Katz v. United States, 389 U.S. 347, 360 (1967).

[3] Katz, 389 U.S. at 351.

[4] Katz, 389 U.S. at 361 (Harland, J., concurring).

[5] L. M. Gladysz, "Status Update: When social media enters the courtroom," 7 I/S:J.L. & Pol'y for Info. Soc'y 691, 2012.

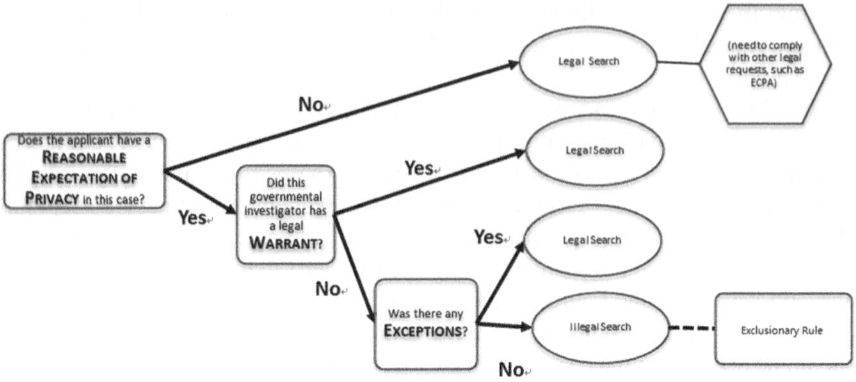

Fig. 1 The structure of a legal search

We can find that this legal system built a privacy button to recognize the private and the public forum, and the trigger is the "reasonable expectation of privacy." While the privacy switches on, we have a reasonable expectation that government will leave its hands away from this private sphere. But the government will conduct a search in the area where the privacy switches off. Generally the government should not put its hands in the private sphere. However, according to Fig. 1, it is not true that we will keep our privacy if we just have a reasonable expectation of privacy. The court made more spaces for government to investigate crimes, especially those exceptional searches without warrants. There must be one more area between private and the public.

Doing searching and collecting evidence on Facebook as the new type search, the legal investigation, however, generally conducts as the same with traditionally searching objects. Scholars coined three levels of privacy in information: public communications, private communications, and quasi-private communications. Communications in public area will not be protected by the Fourth Amendment, because they are open to share. The court held that, if a person voluntarily provided information to others, there is no expectation of privacy in this case. On the other hand, communications in a private sphere means that information is only accessed by a limited group of people, such as e-mail, phone calls, chatting, or other instant messaging services. They are not intentionally divulged to the public, and the court is likely to consider the protection of privacy under the Fourth Amendment in these communications. The level of privacy between the public and private communications is quasi-private communications, which are most difficult to classify. Take communications on Facebook as an example: user chooses to share information with friends, friends of friends, or other user-created networks, and communications cannot be accessible to the general community of Facebook, but are accessible to user's friends or networks, which are called quasi-private communications. Although the information is not available to the general public, judges and legislators still think quasi-private communications as public, because a potentially large number of people will be able to access this information. Users may believe that, on the contrary, they have a reasonable expectation of privacy based on the privacy setting not allowing the public to access their Facebook.[6] Now there is a gap between the legal system and social network site users.

3 Privacy Risks on Social Networking Sites and Its Privacy Protection

To find the reason of forming that gap, privacy risks on social networking sites and mechanism for protecting privacy on them should be considered. Risks currently existing on social networking sites not only challenge the existing protections for privacy, but also reflect the user's attitude of privacy and its protection, which he/she knew and is seeking for.

Privacy Risks on Social Networking Sites

According to the literatures,[7] we can broadly categorize the privacy risks on social network sites into three, which are improper personal information flow, malicious friends, and monitoring and tracking.

Improper Personal Information Flow

For example, a person originally shares his/her information (e.g., disease, treatment) with people who are under the same healthy straits, but due to the nature of social networking sites information is released by chance (no matter legally or illegally) and everyone can possibly access his/her information.[8] People share information on social networking sites and also spread others' information. They may print out, reuse, edit, disseminate, or transmit to others. The court allowed these secondary uses, and stated that there is no invasion of privacy, just because information was disclosure (in the word of the Katz case, "knowingly exposed"). Or like using the tag function on social networking sites, people may tag others on photos posted on his/her Facebook. It not only spreads others' images and appearances, but also

[6] E. W. Sholl, "Exhibit Facebook: the discoverability and admissibility of social media evidence," 16 Tul. J. Tech. & Intell. Prop. 207, Fall 2013, pp. 211–213.

[7] J. Grimmelmann, "Saving Facebook" 94 Iowa L. Rev. 1137, 2009.

[8] J. Grimmelmann, "Saving Facebook" 94 Iowa L. Rev. 1137, 2009, pp. 1152–1153.

identifies them visible for the public. These easy-to-use functions may not need the consents from information subjects, and spread information very quickly and broadly. It is very hard to control this improper personal information flow.[9]

Besides, the third-party application or service providers ask a general authority to access a user's profile, e-mail, friend list, or photos, without specific purposes, when the user wanted to use this application or service on social networking sites. These providers sometimes even ask users' consents to allow them to spread information on users' social networking sites. Users only can choose opt-in or opt-out, and have no stand to choose the degree of disclosure of privacy. It is totally a fake choice to self-determination. The structure, when the social networking sites were designed, has preinstalled these unavoidable privacy risks.

Malicious Friends

Not only from the third-party application or service providers, or the government, the greatest threat to users' privacy on social networking sites is surprisingly from malicious friends. They may improperly reuse, spread, or transmit information, causing information leakage or privacy violations intentionally or recklessly. Building social networks is of course the main purpose to participate in a social networking site; therefore, users will be easy to make a friendship on the Internet with less diligence. And the term "friend" has some magic power to weaken a user's vigilance against infringement of privacy. Then people may unconsciously allow those "friends" easily access and reuse our personal information, but those "friends" are essentially strangers and may cause privacy risks out of control.

Monitoring and Tracking

A user will publish information or open his/her profile to the public, exchanging the chance to make a new friend, in order to create his/her own Internet social community. Additionally, using the "subscribe-to" or "follow-up" function, a user

[9]This paragraph would like to emphasize here that even if the user doesn't tag himself/herself, friends may still track the user; or the user makes the setting of untag himself/herself or removes his/her tag, but friends have still mentioned him/her in the text or description. This is a typical social risk. Facebook users must understand the risks of their own information being spread on the Internet. Moreover, many users, especially elderly and young people, who often do not care or do not understand these privacy settings, use settings by default, that is, being marked. (A 2013 Taiwanese study noted that up to 61% of Facebook users were "not" actively checking privacy settings on a regular basis, and even one in four users did not know that they could adjust their privacy settings. *See* http://www.trendmicro.tw/tw/about-us/newsroom/releases/articles/20131002054038.html (last visited 10 May 2017).) In this regard, the author believes that to reduce the risk, Facebook may be required to change the original settings. However, Facebook will inevitably consider these involving interests and huge business opportunities which information has brought.

will not miss news of his/her friends or interested persons. That is, users on the social networking sites put themselves under someone's monitor and track. This is also an original design of the social networking sites, possibly causing unavoidable privacy risks. For example, the Facebook sidebar ticker visualized users' online activities. It is a convenient tool for tracking someone's activities on social networking sites, such as what he/she posted, what he/she pushed the "like" button, or who he/she just friended, and also allows users to realize that they are watched. This phenomenon gives users some mental pressure such as feeling anxious, self-restraint, or self-censorship, or even further forming an atmosphere of social control.[10]

Numerous active users on current social networking sites and many of them are willing to disclose personal information on the Internet, even though there is a lot of privacy risks existing. From a psychological point of view, users can pursue the identification in a small group or the society, through self-expression on these platforms, such as posting opinions or photos, making comments, and sharing preferences. By self-disclosure of information, users can create a relationship in this virtual world or related to the real world, and they further build a social status on the platforms or connect to the real world. Users easily trend to expose their privacy or underestimate the privacy risk, in order to satisfy the desire for self-expression.[11]

Another reason why users ignore privacy risks on the social networking sites is that users usually evaluate privacy risks wrongly. When most people think about privacy risks, they are most likely to assess risks with various simplified rules of thumb, instead coming from rational. For example, users easily believe that Facebook is a safe social networking site, when most of their friends have Facebook accounts. They may have the illusion of intimacy while they build the friendship on the social networking sites. If considering the bubble effect[12] and the echo chamber[13] as the result of algorithm used on the social networking sites to disseminate information, users may be under illusion that "friends" on their social networking sites share similar opinion and will follow the same standard to protect their privacy. That would lead users miscalculate risks of information arbitrarily spread by others.

The privacy protection mechanism on social networking sites may be the third reason causing users' miscalculation of privacy risks. While a user sets his/her privacy preference limited to share information with the group of friends, he/she thinks the information is only accessible for friends, and may believe he/she can control the flow of information. However, he/she can neither control how his/her friends use or transmit information, nor control information flow exposed

[10]D. J. Solove, The Future of Reputation: Gossip, Rumor, and Privacy on the Internet, Yale University Press, 2008, p. 108.

[11]P. Wallace, The Psychology of the Internet, new edition, Cambridge University Press, 2001, pp. 28–37.

[12]E. Praiser, The Filter Bubble: What the Internet Is Hiding from You, Penguin, 2011.

[13]E. Bakshy, S. Messing, L. Adamic, "Exposure to ideologically diverse news and opinion on Facebook", Science, 10, 1126–1160, 2015.

unlimitedly. For example, the police registered a Facebook account, and pretended a friend with A, who didn't change his privacy settings on Facebook and put them by default. Then this policeman, a fake friend, can access A's friend list, digging his friendship with B, and disclose that relationship A or B didn't want to be disclosed.[14] Users may not be aware or cannot understand clearly how easy the personal information is spread and abused, even though they set their privacy preference, trying to avoid privacy risks on social networking sites. Like while using the third-party application or service, users neglect or willfully disregard risks that broadly authorize the third party to access their information on a social networking account. In fact, it is precisely what social networking sites design, which allows the third party introduce an opt-in-or-no-use choice and force users facing a dilemma in privacy protection and service use.[15]

Users' Thinking Mode About Privacy Risk

Now we can point out the user approaches to evaluate these privacy risks. First, the evaluation is not objective. Users make a judgment based on their past experiences, considering existing interactive models in the real world, to evaluate these risks that happened or may happen. For example, a user decided to only share his/her posts to a group of "friends" on a social networking site. He/she may connect this concept of "friend" to the experience that he/she makes a real friend in the society, and neglect that a social networking site friend may be a person never met or, the worse, a virtual character.

Second, it is users' subjective capacity to control the personal information flow that leads whether users feel their privacy was infringed.[16] Users believe that they can completely control their information. Under different situations, users will partly release information or totally relinquish their control on information, such

[14]D. J. Solove, The Future of Reputation: Gossip, Rumor, and Privacy on the Internet, Yale University Press, 2008. In fact, facing these continuous privacy disputes, Facebook also gradually changes user permissions of privacy settings. For example, now the Facebook user can hide the friend list. But the risk still exists; although users can only see each other that have intersection after hiding the friend list, it's possible to push back a friend's social circle through these mutual friends' friend lists, which were set by default.

[15]The user may have the right to authorize the specific third party to access personal information, but, in the author's point of view, once the user wants to subscribe the third-party service, the user must allow this third party to access part of his/her information, such as personal file, which may lead the user to bear risks of big data. If the user doesn't allow the third party to access his/her personal file with considerations, he/she will undertake the risks that who cannot use this service. This all-or-nothing privacy setting is what this chapter wants to challenge. The author currently has a rough idea, that is, perhaps the third-party service can provide users the alternative of pay for service, if users want to keep their personal information.

[16]D. J. Solove, The Future of Reputation: Gossip, Rumor, and Privacy on the Internet, Yale University Press, 2008.

as a statement noted in a post, "welcome to reprint." Once they found they lost their control on information, they must feel their privacy was infringed. Take the Facebook sidebar ticker as an example: it just combines other original functions, as Facebook claimed, "Ticker shows you the things you can already see on Facebook, but in real time. Keep up with the latest news as it happens. Listen to music with your friends. Click or hover over a story to join in the conversation."[17] It makes users' activities visible on Facebook in real time. Users cannot cancel this function, but can hide or resize it, which only changes this sidebar on a user's computer screen, but other users still can watch this user's activities. Thus, it raised a privacy issue. More precisely, users found that they cannot decide to hide their activities on Facebook, because the ticker collects their activities and announces this information to other Facebook members. They lose their control on this information and feel their privacy is infringed.

Third, when a user wants to use applications or services provided by the third party, he/she will neglect or willfully disregard privacy risks, and authorize the third party to access their account with a general consent. Social networking sites or third-party application or service providers often require users to provide a general consent to allow them to freely access information about users for a commercial purpose and/or an uncertain purpose for the future interest. Users are forced to face this in-or-nothing dilemma, and must disregard privacy risks, in order to access these Internet services.

In sum, when users need to evaluate privacy risks on the social networking sites, they will consider a lot, such as their purposes to use the social networking sites, the relationship among them, and nature and types of information shared. There will be a whole background story behind users' decision to share information. It is a contextual consideration, rather than an open/not open to the public decision.

Privacy Protection from Social Networking Sit Providers

Privacy Policy

The privacy protection offered by social networking site providers starts with the privacy policy. When a person registers in the user account using a social networking site, he/she is preset to have read the terms of the privacy and understood its contents. As Facebook's term of service states, users own contents and information they post on Facebook. Users can manage their way to share contents by privacy settings; set privacy preference; freely post, remove, or modify information on their Facebook account; and also freely decide to deactivate or delete their account. In addition, Facebook also ensures that they will share user information with third parties only in limited circumstances, and will take cautious approach to protect

[17]https://www.facebook.com/help/255898821192992/ last visited 10 May 2017).

users' information.[18] In other words, this approach is based on the assumption of a rational person. Social networking sites have obligations to clearly tell users the use of their personal information. As long as the privacy policy is clearly and honestly released to users, they will have the ability to make appropriate decisions with sufficient information notified in advance.

However, users may not really fully understand their right written in the privacy policy. It is obvious that many users didn't read the privacy policy and just put on the "agree" button, sending their general consent. Even though some users read privacy policy while they register in, they might not fully understand its contents. In Facebook it is stated in its term of service, "we try to keep Facebook up, bug-free, and safe, but you use it at your risk .... We do not guarantee that Facebook will always be safe, secure or error-free or that Facebook will always function without disruptions, delays or imperfections. Facebook is not responsible for the actions, content, information, or data of the third partied, and you release to us, our directors, officers, employees, and agents from any claims and damages, known and unknown, arising out of or in any way connected with any claim you have against such third party."[19] That is, the true meaning of this privacy policy is that the social networking site will not take any response for these possible privacy risks that may happen when users authorize the third-party application or service providers to access users' information on the social networking site. Even this third party may need a permission or payment to be introduced in this social networking site. In the meantime, Facebook's term of service only bound itself; that is, each privacy policy can only regulate this specific social networking site itself. Other user members, third parties, or government investigators will not be regulated by a specific privacy policy on the social networking sites. In this approach, a user general consent or authority is the button to open or close the wall to protect privacy on the social networking sites.

Privacy Settings

Social networking sites introduced a technology model of privacy settings, allowing users to set their privacy preferences. It is also based on aforementioned rational person model. Through the privacy settings, users can decide accessibility of information in a group of people.[20] For instance (Fig. 2), when a Facebook user

---

[18]https://www.facebook.com/privacy/explanation (last visited 10 May 2017).

[19]https://www.facebook.com/privacy/explanation (last visited 10 May 2017).

[20]Specifically, the user can share his/her information through personalizing his/her friend circle by privacy settings. The user can open posts to the public, share information with all friends in his/her friend list, only authorize some friends to access or exclude specific friends to know, include and exclude friends and lists by customization, or even allow no one but himself/herself to see. Users can use their own preferences to decide who can access their information, post, or other activities. These preferences can be set in a unified setting, or users can set them case by case. But the dissemination of information is still the same as indicated in Figure 2, which means, when a user

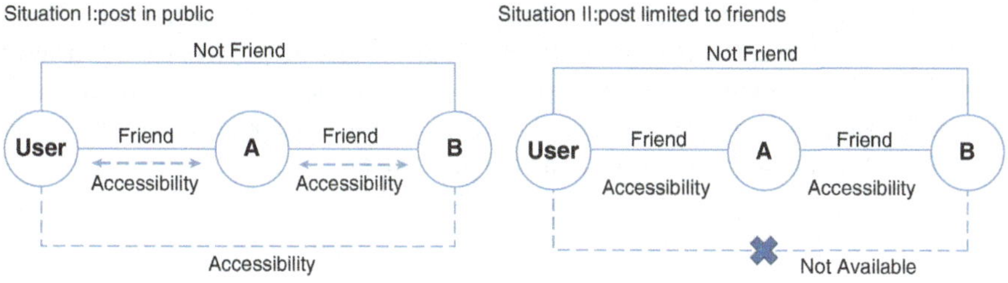

Fig. 2 Illustration of FB privacy setting

sets his/her information in public or uses the privacy setting by default, B, who is not the user's friend, can also see what the user posts on his/her own Facebook page. But, if the user limited to share his/her information in the group of friend, B in theory cannot access the user's posts or photos. Such an operation restates what the social networking site claimed in its privacy policy; users can control personal information and have choices to decide their privacy settings.

However, most users either have no knowledge about this privacy setting interface to set their privacy preferences or have no willingness to change the default privacy setting. According to a statics report, 25% of Facebook users do not use any type of privacy controls.[21] The default setting of privacy mostly is public, because the mission of a social networking site is "to make the world more open and connected."[22] We need to push more on education and promotion of the use of privacy setting, and then this technical protection will be truly played.

In summary, facing multiple privacy risks on the social networking sites, we can find that the approach that users use to evaluate privacy risks is not all-or-nothing, but an approach to make this decision in context. For a user, he/she may make different decisions about the same kind of information but under different circumstances, while privacy risks were raised. Although privacy risks on social networking sites actually have different contexts, the privacy protection mechanism on these sites still inherited the existing decision-making model of general privacy risks, which is, an appropriate access to information comes along with a user's consent or authority. This model requires a person to fully understand privacy risks and to have the ability to make diligent decision. The utility of this model will be reduced in dealing with privacy risks on social networking sites. Users only can accept the privacy policy to join the social networking site and bear the risks of

---

sets his/her preference to public, anyone can access his/her public contents; if the user limits the distribution of the contents, only qualified friends have access to the information. Even a friend of a friend, he/she cannot access these limited contents because of the qualification.

[21] S. Bennet, "Facebook, Twitter, Pinterest, Instagram—Social Media Statistics and facts 2012" All Twitter, http://www.adweek.com/socialtimes/social-media-stats-2012/472135. (last visited 10 May 2017).

[22] https://www.facebook.com/about/basics/ (last visited 10 May 2017).

invasion of privacy, or they give up participating activities on the social networking site, in order to keep their privacy. In this Internet age, actually users don't have any choice. Through the powerful social networks and the Internet, no one can lock his/her privacy in a safe case, unless the whole Internet disappears from the world or he/she stays alone far away from the human society.

4 Legal Safeguards Continuously Narrow Down

Although the privacy protection mechanism on social networking sites is unsatisfactory, scholars combined this privacy setting with the reasonable expectation of privacy standard, and stated that we can use "privacy policy" and "privacy settings" as factors in the application of Justice Harlan's second rule, "the expectation is one that society is prepared to recognize as 'reasonable'."[23] In other words, When a user makes a privacy setting, limited information only to be shared in a small group, he/she has not only a subjective expectation of privacy, but also an objective one. The privacy setting presents a person's will to remain private on the social networking site, and this mechanism can be recognized as a reasonable privacy protection by the society. Therefore, using the privacy setting may secure some private space on the social networking sites. When the information is marked as public, anyone on or off Facebook can access the information, including the government; but when the information is not open to the public, the government needs to obtain it through other means, such as requesting information from social networking sites or requesting information under user's consent.[24] However, there are two cases gradually piercing this mechanism of privacy protection.

Requesting Information from Social Networking Sites

The privacy setting on social networking sites is public by default. In People v. Harris, the defendant was charged with disorderly conduct for marching on the roadway of the bridge when he participated in a protest march on the Brooklyn Bridge.[25] The District Attorney's office requested Twitter to submit the whole Harris's Twitter records through a subpoena, in order to contradict Harris's defense at trial. After receiving a notification from Twitter, Harris moved to quash the

[23]E. W. Sholl, "Exhibit facebook: the discoverability and admissibility of social media evidence," 16 Tul. J. Tech. & Intell. Prop. 207, Fall 2013; Bradley R. Johnson, Untagging Ourselves: Facebook and the Law in the Virtual panoptic on. 13 T.M. Cooley J. Prac. & Clin L. **185** (2011).

[24]In practice, there are more approaches for the government requesting information, such as requests from non-US law enforcement or emergency requests from law enforcement. *See* J. Brunty & K. Helenek, Social media investigation for law enforcement, Elsevier, 2013, pp. 77–85.

[25]People v. Harris, 945 N.Y.S.2d 505(Crim. Ct. 2012).

subpoena based on the privacy right guaranteed by the Fourth Amendment, and the rules set forth in the Stored Communications Act.[26] Twitter also stated that it would not comply with the subpoena until the motion to quash was ruled on.

In an April 2012 hearing, the court denied Harris's motion based on his lacking of the proprietary interest, and compelled Twitter to comply with the subpoena. The court reasoned that, first, the defendant does not have standing to quash a subpoena issued to the third party. Second, according to Twitter's terms of service, Harris has authorized Twitter a license to use, display, and distribute his tweets to anyone and for any purpose it may have, when he signed up to use Twitter's services. Furthermore, Twitter also informs its users that tweets may be viewed by the entire world through the default setting. Thus, the court held that Harris had no proprietary interests in his tweets, because he allowed his tweets to be viewed publicly and anyone with Internet could view, print out, or use these tweets in any way.

Twitter then moved to quash the subpoena itself and refused to comply with this court order. It argued that the terms of service were changed and a section states "you retain your right to any content you submit, post or display on or through the service," which gives Harris the proprietary interests in his tweets. Therefore, denying the defendant standing to quash the subpoena places a burden on Twitter to comply with all subpoenas or to move to quash them on behalf of all defendants. The court denied Twitter's motion in a June 2012 hearing, and found that every third-party service bears this burden and there is no reason making an exception for Twitter. Besides, even though the Warshak court recognized that a person has a reasonable expectation of privacy in his e-mails,[27] tweets in Harris case posted on the Internet as a public forum could not apply the Warshak rule only for private communications. As a result, the court ultimately decided that Twitter needs to comply to the subpoena.

We need to point out here that the court is still applying the public and private dichotomy to determine whether this investigational conduct in dispute infringed the defendant's privacy. The premise in this case is that the nature of tweets is a public broadcast. Since the information itself is open to public, the user's privacy won't be considered further, when law enforcement agencies access the information and collect it as evidence.

---

[26] 18 U.S.C. §§ 2701–2712 (2006).

[27] United States v. Warshak, 631 F.3d 266 (sixth Cir. 2010). "It is the first case from the United States Circuit Court of Appeals to explicitly hold that there is a reasonable expectation of privacy in the content of e-mails stored on third party servers and that the content of these emails is subject to Fourth Amendment protection." *See* https://en.wikipedia.org/wiki/United_States_v._Warshak (last visited 10 May 2017).

Requesting Information Under User's Consent

Even though the user sets the privacy to limit who can access his/her information, the case of United States v. Meregildo, one of the first reported cases to deal with social networking sites and the Fourth Amendment, still pierced this protection on social networking sites. The defendant adjusted the privacy setting on hid Facebook and only allowed his Facebook friends to access his postings. In this case, the government found a cooperating witness, who happened to be the Facebook friend with the defendant, and obtained evidence against the defendant from his Facebook through this friend. Then the defendant moved to suppress the evidence seized from his Facebook account, based on the violation of his Fourth Amendment rights.

The court found: "Where Facebook privacy settings allow viewership of postings by 'friends,' the Government may access them through a cooperating witness who is a 'friend' without violating the Fourth Amendment .... While [the defendant] undoubtedly believed that his Facebook profile would not be shared with law enforcement, he had no justifiable expectation that his 'friends' would keep his profile private. And the wider his circle of 'friends,' the more likely [the defendant's] posts would be viewed by someone he never expected to see them. [The defendant's] legitimate expectation of privacy ended when he disseminated posts to his 'friends' because those 'friends' were free to use the information however they wanted—including sharing it with the Government."[28]

We need to point out, "the Meregildo opinion should not be interpreted to mean that the court is never willing to apply the Fourth Amendment protections to information placed on social networking sites."[29] Moreover, scholars argue, "if an investigator is seeking information about a social media user who has provided consent for the investigator to access or obtain the use's account information, the user should be directed by the investigator to obtain that information on their own from their account."[30]

Discussion

In Harris case, the court held that a social networking site is a public forum, and stated that the defendant has no expectation of privacy on his tweets, because Twitter is a public forum to spread information, and once tweets were knowingly exposed the defendant no longer kept them as private. (In this case, the court decided that the defendant had no proprietary interest in his Twitter account information

[28]United States v. Meregildo, No. 11 Cr. 576(WHP), 2012 WL 3264501 (S.D.N.Y. Aug. 10, 2012).
[29]*See* T. A. Hoffmeister, Social Media in the courtroom- A New Era for Criminal Justice?, p. 77.
[30]*See* J. Brunty, & K. Helenek, Social Media Investigation for Law Enforcement, p. 81.

or tweets.)[31] But in Meregildo case, the court did not clearly deny the defendant expectation of privacy based on his privacy setting. Instead, the court stated that, because the defendant had no justifiable expectation that his friends would keep his profile private, the defendant cannot suppress the evidence seized from his Facebook account and cannot argue the government infringe his Fourth Amendment rights. The court thought that the defendant's legitimate expectation of privacy ended when he disseminated posts to his friends, which means the court still follows the rule built by the Katz case, using whether information was "knowingly exposed" as the standard. If information was knowingly exposed, then there is no expectation of privacy existing.

The court in Meregildo case also implies that this judgment inherited the False Friend Doctrine founded in Hoffa case. The American Supreme Court stated in Hoffa case that there is no infringement of the defendant's constitutional rights, that a criminal investigator, hiding his identity, had a friendship with the defendant, earned his trust, and obtained his confession of issued crime. Therefore, the defendant cannot argue this investigation infringe his Fourth Amendment rights, while he mistrusted this undercover investigator. The court held two reasons. First, the Fourth Amendment does not protect a wrongdoer's misplaced belief that a person to whom he voluntarily confides his wrong doing will not reveal it. Second, as a speaking might be tapped, actions like betray or deception are preexisting risks of human society.[32]

However, the first and the biggest criticism of the False Friend Doctrine is to trust that a friend will not and should not constitute legitimate grounds to exclude privacy. The Fourth Amendment wants to protect people's property and private space from invasion of the government. The key point should be whether this governmental investigation comply these constitutional requests, especially a warrant for searching private area. Besides, this society doesn't ask people to bear every risks of betray. Legislators regulated fraud as a crime and breach of trust as a tort. We can remedy losses caused from these risks through litigation. Specially thinking of the nature and privacy risks on the social networking sites, the trust among Facebook friends is obviously different from the friendship in Hoffa case (they physically knew each other for a period). It is really ridiculous evaluating privacy protection by the number of Facebook friends. We may think that, if a defendant always accepts others as friend without a segment and he/she may have thousands of Facebook friends, the number of friends might become one of the factors to prove his/her lack of objective expectation of privacy. We need to apply factors in context in order to find the appropriate line for privacy protection, especially on social networking sites.

[31] People v. Harris, 945 N.Y.S.2d 505(Crim. Ct. 2012).

[32] "Neither this Court nor any member of it has ever expressed the view that the Fourth Amendment protects a wrongdoer's misplaced belief that a person to whom he voluntarily confides his wrong doing will not reveal it." *See* Hoffa v. United States, 385 U.S. 293 (1966).

Second, from the point of view of protection of personal information privacy, a general consent cannot justify the government freely and arbitrarily access personal information without a re-consent. It is not true that a person will lose the whole control of his/her information, while he/she gives others an authority to access in the very beginning. This person still has the right to keep his/her privacy from information abuse or information use out of the range of original authority. We need to be careful on the chilling effect, especially in Meregildo. If we lose our privacy because of a virtual friend, we might stop to say anything or to make friends on social networking sites, which is not the purpose of a social networking site.

Third, in Meregildo, the court stated "[the defendant] had no justifiable expectation that his 'friends' would keep his profile private." This is the issue of risk acceptance, who need to bear this risk. We all need to bear some risks in society, such as there will be the third party tapping the conversation while we are talking loudly on the sidewalk. But it does not equal to making the consent. Consent will justify a warrantless government investigation and clearly show the range of privacy protection. Therefore, using information to charge one's crime is out of the range of information sharing on the Facebook. It is hard to justify transmission of information with the defendant's consent.

In sum, the purpose of the Fourth Amendment cannot be reached only by using this on/off privacy button. The simple statement, whether a person has privacy, cannot keep the arbitrary governmental searches away from our private social networking site. We need to find a more effective rule to protect our privacy on the websites.

5 Try to Find the Solution

Now we know that whether the privacy had been violated on the social networking sites should be discussed in context. From the users' perspective, they will make different arguments on privacy related to the same photo different situations. For example, in the Burning Man Festival, participants can tolerate or even are willing to allow their pictures to be taken by other participants, and to be shared/to share these photos in their group on the social networking sites. But these participants will feel that their privacy has been infringed while these photos are reproduced in newspapers or magazines or their websites. This is privacy in context. Privacy protection is not an on/off button, but it should be considered under the whole background of this event and related social norms. Especially in cases related to a criminal search on the social networking sites, it is necessary to consider the context of every single case, in order to judge whether a governmental investigation infringed the user's privacy on the social networking site. Here is Professor Helen Nissenbaum's privacy in context theory, and we tried to use this theory as the solution to solve problems.

"Privacy as Context" Theory

In Nissenbaum's theory,[33] protection of personal information should connect with personal information flow in context. She provides a useful and clear benchmark to determine whether the information is worth protecting or privacy has been infringed, which not only does focus on the appearance of information at the single point, but also takes the information as the flow and considers the overall movement of information. When information transmits from one "information subject" to another, there should be different norms for information corresponding to different contexts of the case. A context broadly includes situations for each personal life, such as educational, political, economic, and social dimensions, and abstractly means experiences on a variety of social structures in everyone's daily life. It is influenced by culture and history, natural environment, political structure, economic systems, religious institutions, and so on in a specific time and space in a specific society. Any context will follow the corresponding norms, and these norms will decide for individuals what the role to play, and what actions to take or be not taken in certain contexts. Norms for information are developed by relevant agents, type of information, and transmission principles; and relevant agents include information subject, information owner and information deliver (they might be or might not be the information subject), and information recipient. Each information flow will follow the corresponding social norms. For example, the airline can ask the destination of the passenger for sure, but it is inappropriate that they ask the passenger what they will do there, whom they will meet with, or where they will spend the night. These social norms are used to determine whether the privacy in this case is worthy of protection.

This theory jumps out from public/nonpublic dichotomy, and in discussing the privacy issue on personal information it is argued that full account of the context of the situation should be taken, and observed as a whole.

Response to the Previous Question

We can find two questions from previous cases. First, in Harris case, who is the owner of those disputed tweets and profiles? In what situation the information subject is allowed to disclosure? If the disclosure is allowed, what part should be disclosed? Second, in Meregildo case, is it allowed that a Facebook friend reveal the defendant's information to the police?

The first question involved background story of this case, and we can use "appropriateness" as the standard to make a judgment. "Appropriateness" in this theory can be used to clarify in a specific context what personal information is

[33] H. Nieenbaum, Privacy in Context: Technology Policy and the Integrity of Social Life, Stanford Law Books, 2009.

allowed to be disclosed, what personal information is expected to be disclosed, or what personal information even should be disclosed.[34] In Harris case, the defendant created these tweets and profiles and thus got a position of information subjecte and information owner. Twitter is a media, providing a platform for users spreading their information. Twitter holds copies of these tweets and profiles, just because holding information is necessary in order to provide services. It is not appropriate taking Twitter as the information subject or information owner under this context. Besides, as Twitter mentioned in the second hearing, it changed the terms of policy to clarify that users are information owners. In Harris case, the District Attorney's office acquired Harris's Twitter records through a subpoena in order to contradict Harris's anticipated defense at trail. The District Attorney's office used a subpoena to Twitter, instead of a warrant to Harris who is the right person to decide this personal information. They tried to avoid privacy protection guaranteed by the Fourth Amendment. Since the District Attorney's office wants to obtain the contents of Harris's tweets, according to the Fourth Amendment, they need to apply a warrant issued by the court, and explain the probable cause and the range of their search in Harris's Twitter account.

The second question involved the issues on information transmission, which can be solved through transmission principles in this theory. The transmission principle is norms regulating conditions and limitations of transmitting information, in order to decide whether this transmission of information should happen under this situation. In other words, the transmission principle will limit transmissions of information among agents, according to requests in the context. For example, if a confidentiality requirement was signed between information subject and information recipient, the recipient is not allowed to transmit information from the subject to others. The recipient may have appropriate position to reuse this information, but he/she is not allowed to do so, based on the confidentiality requirement, which is the transmission principle in this case.[35] In Meregildo case, we can understand in a general common sense that the defendant allowed Facebook friends to access his posts and photos, which does not mean he gave a broad authority to these friends to use his information arbitrarily. If he knew a friend will transmit his information to the government, the defendant definitely prohibits this friend to access this information, or even unfriends him. In a general social situation, when we want to obtain friends' photos or posts, we will notice our friends and tell them the purposes and range the information will be used. Although there is no expressed provision in this situation, such a norm is generally respected in the society. Therefore, transmission of information in Meregildo case breaches such a social norm regulating information flow in the context of this situation.

[34] H. Nieenbaum, Privacy in Context: Technology Policy and the Integrity of Social Life, Stanford Law Books, 2009, p. 138.

[35] H. Nieenbaum, Privacy in Context: Technology Policy and the Integrity of Social Life, Stanford Law Books, 2009, pp. 145–147.

Discussion

Rather than beyond the reasonable expectation of privacy standard, this privacy as context theory is as a supplement to interpret the reasonableness in Katz case. Because this theory provided social norms in context to explain what people will do in this situation, that is, the expectation is one "that society is prepared to recognize as 'reasonable'." Thus, this theory gave the new legal concept under the past norm, and created a possible interface to apply the old standard to issues, arriving at a more effective reference to protect privacy on the social networking sites. This theory is still building, such as the transmission principle has to be formed by cases. But the court can analyze and argue different situations case by case, and further form detail specifications of context, in order to provide more strength safeguard for privacy on the social networking sites. Then let users fully appreciate what is "to make the world more open and connected."

6 Conclusion

Justice Harlan provided a reasonable expectation of privacy as the line to prevent the government invasion of people's private sphere, but the legal system in past cases stiffly operates this standard as nothing but an on/off privacy button. Under emphasized considerations for interests of prosecuting crimes, people easily lose their privacy, especially the privacy on social networking sites. Because users need to make a general consent while they create an account, applying with the privacy button, what they post or disclose on the social networking site almost all are treated as public, and then the government would arbitrarily search for criminal evidence without following the requirements under the Fourth Amendment. Even though the user set his privacy preference and limited his information to a small group of friends, this privacy protection was pierced by the Meregildo case, in which the court admitted that the user needs to bear the risk of malicious friends. Then the court opened this quasi-private area to the government, and broke the privacy protection under the Fourth Amendment. The court didn't say that there is no expectation of privacy in this area, but allowed the government to use other means to avoid privacy protection. Now we can find this legal model to protect privacy and doesn't consider privacy risks on social networking sites and users' thinking mode about privacy risks. Since users consider their disclosure and evaluate the privacy risks in context, the privacy in context approach should be considered while we are thinking of the protection of privacy. Nissenbaum's theory can be a knob of information disclosure and transmission, adjusting to apply in different situations case by case. It is more effective to protect privacy on social networking sites. It still follows the reasonable expectation of privacy standard, at the same time, and further resupplies the objective reasonable test.

References

1. B.R. Johnson, Untagging ourselves: facebook and the law in the virtual panopticon. 13 TM Cooley J. Prac. & Clin. L. **185**, (2011).
2. D.J. Solove, *The future of reputation: gossip, rumor, and privacy on the internet* (Yale University Press, 2008).
3. E. Bakshy, S. Messing, L. Adamic, Exposure to ideologically diverse news and opinion on Facebook. Science. **10**, 1126–1160 (2015)
4. E. Praiser, *The filter bubble: what the internet is hiding from you* (Penguin, 2011).
5. E.W. Sholl, Exhibit facebook: the discoverability and admissibility of social media evidence. 16 Tul. J. Tech. & Intell. Prop. **207**, (2013).
6. H. Nieenbaum, *Privacy in context: technology policy and the integrity of social life* (Stanford Law Books, 2009).
7. J. Brunty, K. Helenek, *Social media investigation for law enforcement* (Elsevier, 2013).
8. J. Grimmelmann, Saving Facebook, 94. Iowa L. Rev. **1137**, 52–59 (2009)
9. L.M. Gladysz, Status update: when social media enters the courtroom. 7 I/S: J.L. & Pol'y for Info. Soc'y **691**, (2012).
10. P. Wallace, *The psychology of the internet*, new edn. (Cambridge University Press, 2001).
11. S. Bennet, "Facebook, Twitter, Pinterest, Instagram—Social Media Statistics and facts 2012" All Twitter, http://www.adweek.com/socialtimes/social-media-stats-2012/472135. Accessed 10 May 2017
12. T.A. Hoffmeister, *Social media in the courtroom- a new era for criminal justice?* (Praeger, 2014).

Emergency Detection and Evacuation Planning Using Social Media

Coşkun Şahin, Jon Rokne, and Reda Alhajj

Abstract Social media platforms have become an important part of our daily lives, especially in the last decade where number of users and messages communicated are tremendously increasing. They are commonly accepted as the easiest and fastest means of sharing recent news with others. Social media users are timely informed about all incidents taking place around the world. This is possible because people tend to share incidents on the spot despite severe consequences in cases like accidents where people in close relation with potential causalities may get shocked. On the other hand, instant communication on the social media may lead to positive impact and immediate benefit. Thus, utilizing social media has a great potential to handle specific situations, like earthquakes, terrorist attacks, and civil disorders, where getting organized timely and efficiently is crucial. In this work, we propose a system that uses strengths of social media to detect emergencies, inform and lead organizations and people so that the loss and damage can be minimized. The proposed model employs multiple agents in the emergency management process. It is capable of analyzing social media posts, filtering irrelevant and unnecessary content, detecting crisis and emergency situations, summarizing them by giving detailed information about the location and impact, creating an evacuation plan where and when possible.

1 Introduction

Recently, the number of natural disasters (like earthquakes, flooding, thunderstorms, volcanic eruptions, etc.) around the world has notably increased. This is mostly attributed to global warming. In addition, human-sourced emergencies including wars, riots, and terror attacks have resulted in loss of lives, injuries, and financial damage. In such crisis and emergency scenarios, organizing and responding timely

---

C. Şahin · J. Rokne · R. Alhajj (✉)
Department of Computer Science, University of Calgary, Calgary, AB, Canada
e-mail: coskun.sahin1@ucalgary.ca; rokne@ucalgary.ca; alhajj@ucalgary.ca

T. Özyer et al. (eds.), *Social Networks and Surveillance for Society*, Lecture Notes in Social Networks, https://doi.org/10.1007/978-3-319-78256-0_9

and efficiently can prevent the situation from elevating worse and it may help to overcome or stop the damages rapidly. In disaster management, timely coordination and information sharing is crucial for emergency response and rescue teams. When multiple organizations and the general public are trying to help in such a situation, the lack of a central platform to communicate and get information will decrease information sharing and thus success rate. Sometimes people try to get the latest news about an event both to find out its effect on their neighborhood and for curiosity.

With the advance in technology and the rise of web 2.0, micro-blogging has become very popular. It is a type of blogging where users can send short status updates which may be accompanied with micromedia like photos and audios. As a result, social media platforms, like Twitter and Facebook, already have clearly visible and huge impact on people's lives. Users share information, news, and their opinions on social media using texts, pictures, and videos. After these social communication platforms became available on mobile devices, the number of active users on social media platforms has increased rapidly. Mobility and easy access make them great sources of real-time, first-hand news about events around the globe. As a result of excessive information sharing and consumption, researchers started using social media as a valuable source for data that could be analyzed for various types of valuable discoveries which may drive decision making. By analyzing public opinions and orientation, researchers have managed to extract valuable information which can be used for both future prediction and past event summarization. Recent studies have conducted sentimental analysis on social media comments for specific events, including election result and stock market prediction, contagious disease detection [1, 2].

The increase in popularity of social media has attracted the attention of official organizations to join the trend. In order to reach more people, government agencies, companies, and news agencies are active on social media in addition to their own websites. Thus, instead of browsing them one by one, people tend to follow on these platforms organizations they find interesting or they are related to. This makes social media both formal and informal source of information, including daily life usage. People tend to be more active on social media during emergencies both for sharing up-to-date news and to get situational awareness. Citizens have turned them into disaster-related, early indicator communication tools [3]. But high volume of content makes them impossible to use for real-time monitoring without an automatized mechanism that will retrieve content and summarize essential parts.

The publicity and anonymity of social media can be both beneficial and hazardous. It increases accessibility and activeness, making it easier to reach more people with more content. But, it can be used by malicious people to spread misinformation, too. During crisis and emergency situations, high level of uncertainty and chaos push people to believe rumors very easily. For instance, during England Riots in August 2011, the claim that wild animals being freed from the zoo caused panic among people. Similarly, the false news of police beating a 16-year-old girl increased the tension and led the situation to go worse [4]. Besides, not all posts are about current emergencies. There are multiple discussions about various topics and irrelevant data should be filtered to collect accurate information about events.

All these reasons show the necessity of having an automatized system that will filter credible content, eliminate redundant data, cluster them based on the event they are related to, and visualize the results to the user. In this work we propose a multiagent based system that is able to detect emergency situations using social media posts with their locations. In addition, further information extraction gives a summary, including number of causalities (i.e., deaths and injuries) and magnitude in case of an earthquake for instance. The system presents the analysis in a user-friendly environment that includes location of the event on a map and an evacuation plan if the specified location has a building plan in our system.

The rest of the paper is organized as follows: Sect. 2 summarizes existing work that has been published related emergency detection. In Sect. 3, the proposed system architecture and its components are discussed. It gives detailed information about the filtering process, location detection approach, information extraction and evacuation planning methods. Finally, Sect. 4 includes a brief summary of the study and the conclusions.

2 Related Work

The advent of social media attracted researchers to collect and analyze real-time and past data for emergency detection and crisis management, e.g., [5]. In this section, we discuss from the literature the most relevant works in emergency detection, credibility computation, location detection, and information extraction.

EAIMS [6] is a crisis tackling and analysis tool that provides automatic identification, tracking, summarization, and management of recent emergency events from social media. It is designed to be used by emergency management organizations. Apart from event detection, location information, credibility estimation and community identification are provided by the system. The authors run sentimental analysis in social media to track public opinions about emergency responses. Event detection approach in EAIMS is based on locality sensitive hashing [7], where textually similar posts are mapped to similar hashes with high probability. For location detection, they use Carmen Twitter Geolocation System [8] and a machine-learned regression model proposed by Castillo et al. [9] for credibility estimation.

Civique [10] is another emergency detection system which mainly targets on geo-tagged tweets. It uses state-of-the-art classification techniques such as Support Vector Machines (SVM) and Naïve Bayes for training. Similarly, Avvenuti et al. [11] built another emergency management system which contains alert dissemination in addition to emergency detection and emergency monitoring components. Their emergency detection module uses a burst detection technique. For the same purpose, Verma et al. [12] preferred Maximum Entropy (MaxEnt) machine-learning classification method, which gave better results than Naïve Bayes. Sakaki et al. [13] proposed a method that classifies tweets based on keywords, tweet length, and their context for emergency detection. Popescu et al. [14] talk about a method that automatically extracts events and entities from Twitter posts.

Distributed artificial intelligence (in particular multiagent) based emergency handling models have also been developed, e.g., [15–19]. For instance, Hawe et al. [15] described an agent based model to help in minimizing time required to arrive at the hospital in case of emergency and the need to evacuate causalities. Bandyopadhyay and Singh [16] developed an agent based system to study, based on some data collected in India, the behavior of fighter fighters, and the way they choose their routes in case of emergency. Based on the outcome, they proposed routes for fire related emergency to help in handling incidents better. Safdari et al. [17] described a multiagent based approach for handling pre-hospital emergency management. Lopez et al. [18] discussed how it is possible to adapt a multiagent based approach to coordinate ambulances for emergency medical services. Mohammadzadeh et al. [19] described a system that utilizes multiple agents to effectively manage emergency cases which involve chronic heart failure.

3 The Proposed System

Despite all the advantages of social media platforms, they have some shortcomings which make the analysis harder. First, there is no standard format for messages, and people mostly express themselves in an unstructured way. Second, text messages are shorter compared to news articles. For example, Twitter limits their users to 140 characters for each message, called tweet. Thus, a system aiming event detection and information extraction from social media should consider these issues and propose solutions to tackle problems that may occur. The system architecture of our multiagent based model is depicted in Fig. 1. It has four main components, which are:

- Crawler
- Core module
- Evacuation planner
- User interface (UI)

Before going into details about the system, there are two crucial issues to point out. The first one is the result of the system should be accurate. Detecting a non-existing emergency or giving false information about a real one may mislead rescue teams and the general public. Thus every component must be as sensitive as possible. Second, the system must run efficiently to process huge amount of data in real time. In order to prevent a component from blocking the other ones, we use a pipelined framework. In this model, components operate independently and the output of the current operation is passed to a queue that will be consumed by the next component.

Fig. 1 Proposed system architecture

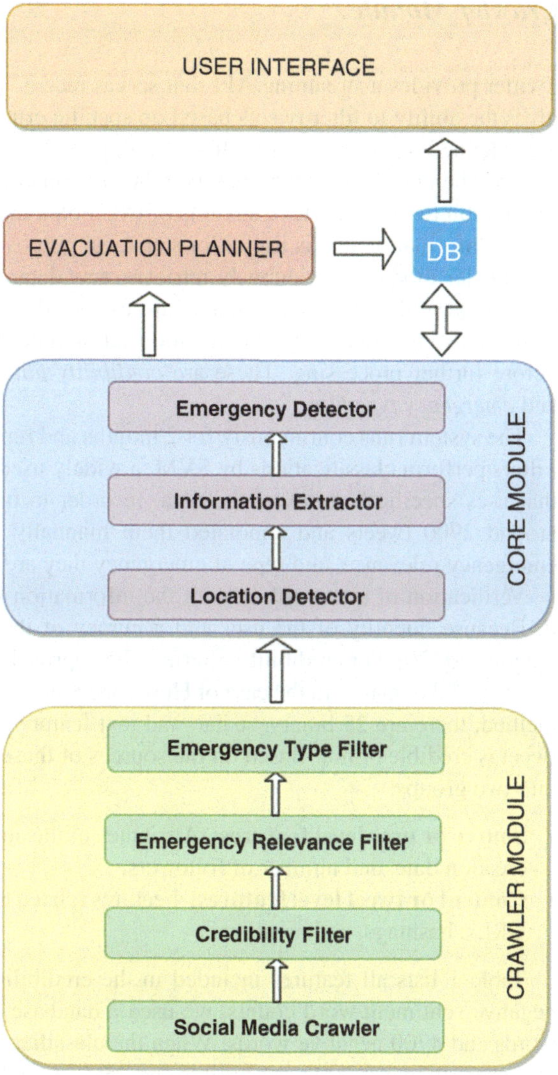

Except for the crawler component and credibility filter, all other components are independent of the social platform that is being crawled. Plugging in another social media crawler, which is able to extract the necessary information, with a customized credibility filter, will not affect the remaining processes. In order to test the system, we choose Twitter as content source of our application.

Crawler Module

Twitter provides a streaming API that serves recent tweets in JSON format. It also gives the ability to filter tweets based on specific criteria. Our crawler module uses a predefined set of keywords, like "earthquake," "fire," "alarm," "explosion," to retrieve the latest tweets that may be related to an emergency, in English language. In addition to the tweet text, streams of tweet data include some metadata and user profile information that is useful for statistical purposes and further analysis.

Crawler module continuously retrieves new data. Except for the retweets, each tweet is parsed and some attributes are extracted in order to be used in the next steps. It also includes three classification submodules to eliminate unnecessary data before further processing. These are *credibility filter*, *emergency relevance filter*, and *emergency type filter*.

The system runs continuously for 2 months and reports emergencies in real time. Filters perform classifications by SVM, a widely used machine learning algorithm, that uses specific features of the data. In order to train the models, we collected around 2900 tweets and annotated them manually in terms of their credibility, emergency relevance, and type of emergency they are related to.

Verification of trustworthiness of the information is necessary before analyzing it. Because, identity of the user and accuracy of the content of a post cannot be guaranteed [20]. For credibility filtering, the approach that Gupta et al. [21] used for detecting fake images in the case of Hurricane Sandy has been implemented. In their method, there are 25 binary, scalar, and text features which are used to decide if a tweet is credible or not. Based on the sources of these features, they can be divided into two groups:

- **Source or user level features:** Attributes of the author, including name, account creation date, and number of followers.
- **Content or tweet level features:** Features related to content of a tweet, including URLs, hashtags, and mentions.

Table 1 lists all features included in the credibility classifier. For positive and negative sentiment word counts, we used a database which contains 2000 positive words and 4700 negative words. When the classifier is trained with the initial data set, where there are 600 non-credible and 2300 credible data, the accuracy is 93.6%, compared to the original authors' 96.65% result with a J48 decision tree classifier. But the results are highly dependent on the domain, the training data size and the content.

Similar to credibility filter, emergency relevance filter is a binary classifier which decides whether a tweet is related to an emergency situation or not. It uses an SVM classifier with a set of predefined keywords. Features of each tweet are determined based on the occurrence of these keywords in the tweet text and hashtags. Our system is able to identify emergency-related posts with an accuracy of 88.4% with the used training data. The output of credibility filter is passed to emergency relevance filter for the analysis.

Table 1 Credibility classification features

User features
Friend count
Follower count
Follower/friend ratio
Number of times listed
User has a URL?
User is verified?
Age of the user account
Tweet features
Length of tweet
Word count
Contains question mark?
Contains exclamation mark?
Question mark count
Exclamation mark count
Contains happy emoticon?
Contains sad emoticon?
Contains first order pronoun?
Contains second order pronoun?
Contains third order pronoun?
Uppercase character count
Negative sentiment word count
Positive sentiment word count
Mention count
Hashtag count
URL count
Retweet count

Emergency type filter is a multiclass classifier which categorizes the data passed by emergency relevance filter. It uses a set of keywords to identify the type of emergency, where SVM model features represent the occurrence of these keywords in tweet text and hashtags. Each tweet that contains information about more than one type of emergency, such as a storm causing flooding, is ignored to avoid ambiguity. There are six types of emergency defined in the system. These are:

- **Earthquake:** Includes all sizes of earthquakes reported on social media.
- **Flooding**
- **Fire**
- **Storm:** Includes tornadoes, typhoons, and hurricanes.
- **Civil Disorder:** Terror attacks, explosions and riots are in this category.
- **Other Emergencies:** Volcano eruptions, snowfalls, car accidents, and radioactive emergency are listed under this type.

With our training data, 87.9% of emergency related tweets are categorized accurately.

Core Module

The core module is responsible for enriching the content of social media posts, clustering them to detect emergencies, finding the locations mentioned in the posts, and extracting additional emergency-specific information including number of deaths, injuries, and earthquake magnitudes. The sub-components performing these tasks are explained in detail below.

Location Detection

In case of emergency management, discovering the geolocation of the posts related to the considered emergency case is a crucial part of the analysis. According to Gonzales et al. [22] and Oussalah et al. [23] geolocation can be identified in four basic ways in Twitter.

- **"Place" object in tweets:** Around 1% of the tweets that Twitter API provides contain location information. Users can tag a specific location or it can be automatically provided using GPS data. Some of this data is in high resolution, including street or building names. Having the *"place"* tag present in a tweet means that Twitter has already performed geolocation, and it provides ready-to-use information.
- **"Coordinates" in tweets:** Some users geotag their tweets with latitude and longitude values based on GPS data. Using some APIs for reverse geotagging, detailed information, such as country and city name, can be extracted from the messages.
- **"Location" in user profile:** Some profiles include location information of the user in a free text format, which can be resolved by using some APIs and dictionaries. But user profile location rarely gives accurate information about user's position while the tweet is being written. When a user travels or provides incorrect information in this field, the data will not be helpful for the analysis.
- **Content-based geolocation:** Some users mention about specific locations in their posts. With high probability, this location information is related to the event that they are talking about. It is possible to extract this information by using some natural language processing (NLP) techniques.

While geo-tagged tweets give accurate information about location, only 2% of them include this data [24]. Dredze et al. [8] used the first three methods in their Carmen geolocation system which is able to resolve locations with 22% success. Carmen provided a hierarchical location structure, which is constructed from Twitter API data. It is publicly available online and has been integrated into our system.

When talking about an emergency situation, people tend to specify event location explicitly. An emergency without a location does not provide useful information to users. They may contain suspicious data which results in false alarms [10]. Thus, posts that do not specify a location are eliminated by the system. In order to utilize the data in hand as efficiently as possible, the system performs content-based geolocation method and combines the result with the one that Carmen produces. It is also necessary to create a system which is independent from the social media platform. All the other three methods are application specific and may not be available in some other platforms.

Our approach for content-based location detection consists of two main steps. Firstly, location entities are extracted from texts using a Named Entity Recognizer (NER) which covers the task of identifying and classifying predefined types of named entities like persons, organizations, locations, temporal and numerical expressions [25]. We use Stanford Named Entity Recognizer (NER) [26] to label word phrases that correspond to location names in tweet messages. The ones tagged as *location* by the NER component are passed to Geonames API for resolving the exact locations with their latitude and longitude attributes.

Information Extraction

Information extraction addresses the problem of finding particular entities, their relationships and events with predefined attributes [25]. Emergency related posts may include additional information about the corresponding event. It can guide rescue teams and give an idea about the volume and effects of an incident. Many information extraction techniques depend on some pre-processing steps, which are:

- **Sentence analyzing and tokenizing:** This is the task of determining the boundaries of sentences and splitting each sentence into tokens.
- **Part of speech (POS) tagging:** In this step, each word is tagged with a grammatical category such as verb, noun, adjective, etc. However, basic grammatical labels are insufficient for detailed analysis. There are some well-known tag sets including Penn Treebank with 45 tags and Brown tag set with 179 tags which are commonly used in available NLP tools.
- **Parsing:** Parsing is the process of splitting words into groups like noun phrases and verb phrases.
- **Dependency analyzing:** This is the task of finding relationships between different entities in a sentence. For instance, the sentence "UofC is located in Calgary." includes *located* relation between *UofC* and *Calgary* entities.

We used Natural Language Toolkit (NLTK) [27] for extracting information from social media posts. NLTK is a set of libraries for natural language processing in English, written in Python. In our system, NLTK performs tokenizing and part of speech tagging before further analysis. It uses Penn Treebank tag set for labeling words, and returns grammatical hierarchy as a tree structure.

The system is capable of finding the number of death or injured persons, as well as the magnitude of an earthquake. By analyzing emergency related messages, it is possible to infer a generic grammatical structure for phrases which include specific information. For instance, after applying POS tagging of NLTK to the sentence "23 people died and 250 injured in Manchester Arena terrorist attack.", the part "23 people died" is tagged with sequence "CD NNS VBD," where in Penn Treebank tag set, *CD* is the cardinal number, *NNS* is the plural noun and *VBD* is the past tense verb. By running multiple tests on posts that include emergency specific information, a regular expression is created by merging POS tagging sequences. This expression is passed to NLTK to produce a custom POS tag. Finally, the tree resulting from the analysis is traversed to find nodes labeled with the custom POS tag.

Emergency Detection

Emergency detection relies on clustering emergency related posts after location tagging and information extraction, based on emergency type and coordinates of the event. With a small margin of error on coordinates, every emergency post is checked to determine whether it could be matched with an existing emergency. After an unsuccessful attempt, a new instance is created. In order to treat an emergency instance as a real emergency, there is a threshold value for the number of posts that should match a specific instance. Posts which cannot pass this threshold are cleared after a certain time for stability concerns. Information extracted in the previous section is attached to the corresponding emergency. For handling multiple posts stating conflicting information related to a specific instance, majority decision combined with trustfulness of the source and dependency between various posts have been adopted.

Evacuation Planner

As a result of development and growth in population, immediate consequences of emergencies are more visible and serious than ever before. This could be realized in terms of number of causalities and the damage directly or indirectly affecting the economy and the environment. The latter destruction may last for longer period, even with ripple effect. To avoid such unpleasant circumstances, it is necessary to predict such scenarios and their possible outcomes in order to get prepared and decrease unwanted consequences. One of the fundamental things that should be practiced before an emergency event is implementing evacuation scenarios to ensure that evacuees will leave the site safely, especially in public places such as theaters, cinemas, hospitals, malls, etc. Unfortunately, whatever is done, it is hard to get all people present at an emergency site educated to follow a certain evacuation plan. The crowd may have diverse background, knowledge, practice, etc. A well-educated person who already participated in evacuation plan training may panic and behave differently when it comes to a real actual emergency.

Simulating evacuations in real-world is a tedious job. As a result, software tools are preferred to design emergency exits and simulate evacuation plans with the current setup. Crowd simulation systems can be grouped into three basic categories [28]:

- **Fluid or particle systems:** These systems rely on the assumption that there is an analogy between crowd behavior and particle motion.
- **Matrix-based systems:** In this type of systems, floor is modeled as a 2D-grid. Each cell can be occupied by an object or a person, or it can be empty.
- **Emergent systems:** Emergent systems aim to utilize interaction of simple parts to simulate crowd behavior.

Multiagent systems help us to simulate real-world environments by using autonomous agents. For crowd behavior analysis, it is a suitable choice as it allows agents to act independently as well as by interacting with each other. There are a wide range of multiagent based commercial tools that are used to design safe egress for places and simulate evacuation scenarios on them. Their aim is to design the environment and agent behavior as realistic as possible to achieve more successful results. Thus, agents differ from each other in terms of physical aspects, such as age, gender, and health, and influence on other people depending on the social hierarchy. Each agent has a limited vision range which is determined by its position and obstacles blocking the way. Psychological factors are also included to be able to simulate common crowd evacuation behavior like herding, queuing, and clogging.

Evacuation planner in our system models each floor as a 2D-grid world where each evacuee is represented as an agent. The aim of this module is not to help in designing better egress or simulate different evacuation scenarios. The main objective is rather to guide people in a way that they can leave from emergency site in minimum time. Thus, each agent can see the whole map, selects the best exit and moves towards it. The process which has been adopted by the evacuation planner is given in Algorithm 1.

While most of the procedures are self-explanatory, there are some points that may require further explanation for the whole process to be better understandable. The distance between two cells is defined as Manhattan distance, since the environment is a 2D-grid. In the second algorithm, assigning an agent to a specific exit gate depends on the load on that exit and the availability of alternatives exit(s) which may be also safe to try. Even distribution of agents on safe exits is preferred to avoid herding problems. During the simulation, in each time-slot, agents can go to north, south, east, west or stay in their current position if the next cell is occupied by another agent. After each agent reaches to the target exit, simulation results are stored in a database to be used by the UI module as discussed next.

Algorithm 1 Evacuation plan finder for a given 2D-grid

```
procedure FIND-EVACUATION-PLAN
    ASSIGN-EXITS
    FIND-EXIT-PATHS
    for i ← 1..n do                                          ▷ iterate over list of agents
        Move agent i starting from the initial position
    end for
end procedure
procedure ASSIGN-EXITS
    Let S be an empty set storing tuples (dist, agent, exit)
    for i ← 1..n do                                          ▷ iterate over list of agents
        for j ← 1..m do                                      ▷ iterate over list of exits
            add (d, i, j) to S, where d is the distance between the agent and exit
        end for
    end for
    Sort S in terms of distance in increasing order
    for j ← 1..s do                                          ▷ iterate over list S
        Let (d, i, j) be the current tuple
        if Exit j is not full and agent i is free then
            Assign agent i to exit j
        end if
    end for
end procedure
procedure FIND-EXIT-PATHS
    for i ← 1..n do                                          ▷ iterate over list of agents
        Find shortest path from agent's position to target exit
        Generate action sequence
    end for
end procedure
```

The UI Module

Visualization of the results is as important as the analysis. Thus, our system contains an easy-to-use user interface which is capable of listing and summarizing emergencies with some additional features. It consists of three main components, namely, emergency list, emergency detail, and evacuation plan.

In the "emergency list" component, users can list the latest active emergencies with the type, location, and date of each emergency. Figure 2 shows an example of this page. It is also possible to search for emergencies in a particular location.

The "emergency detail" component provides additional information about a particular event, including number of death, injuries and magnitude of the incident, e.g., earthquakes. It also lists the latest posts linked with the corresponding incident, as shown in Fig. 3. Location of an emergency is visualized using Google Maps API.

The "evacuation plan" component simulates actions of agents which were created in the evacuation planner submodule. The user can adjust basic visualization features, such as opacity, speed of the simulation, and starting/stopping the simulation. Moreover, it gives the ability to focus on the plans in certain floors of a building by making the remaining ones invisible. A sample evacuation plan visualization is given in Fig. 4.

Type	Location	Start Date
Civil Disorder	CHINA	Jun 15, 2017, 6:17:08 PM
Earthquake	HAWAII	Jun 7, 2017, 4:39:56 AM
Civil Disorder	MAIDUGURI	Jun 7, 2017, 1:59:22 PM
Civil Disorder	SYRIA	Jun 6, 2017, 5:33:43 PM
Civil Disorder	LONDON BRIDGE	Jun 6, 2017, 5:31:59 PM
Civil Disorder	IRAN	Jun 7, 2017, 12:15:07 AM
Civil Disorder	UK	Jun 6, 2017, 10:08:46 PM
Civil Disorder	FLORIDA	Jun 6, 2017, 10:03:09 PM
Civil Disorder	US	Jun 6, 2017, 8:37:08 PM
Earthquake	MYANMAR	Jun 8, 2017, 3:57:41 AM

Fig. 2 Emergency list component

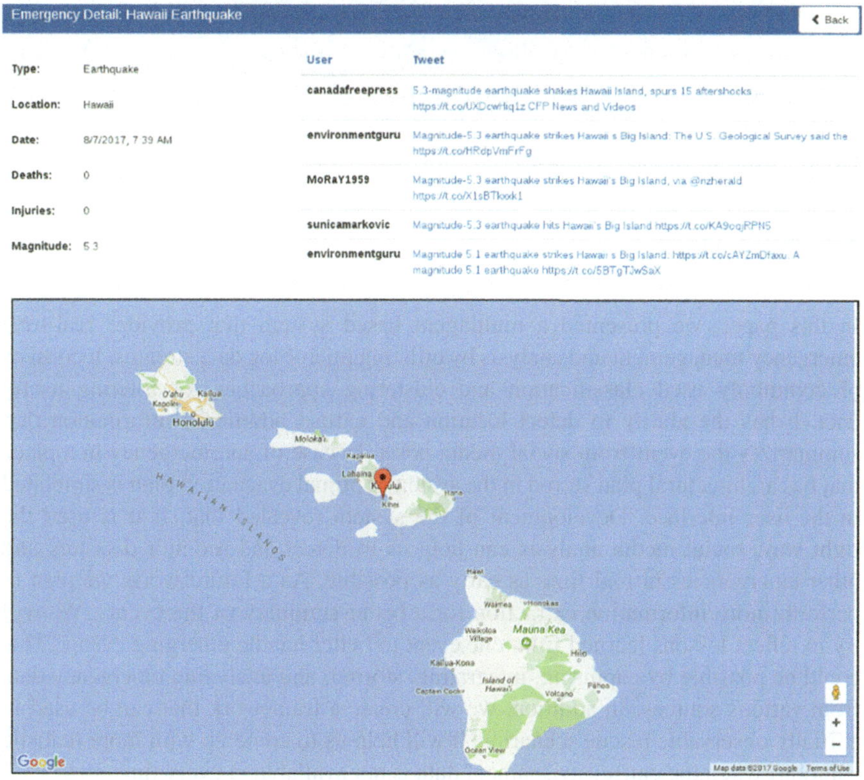

Fig. 3 Emergency detail component

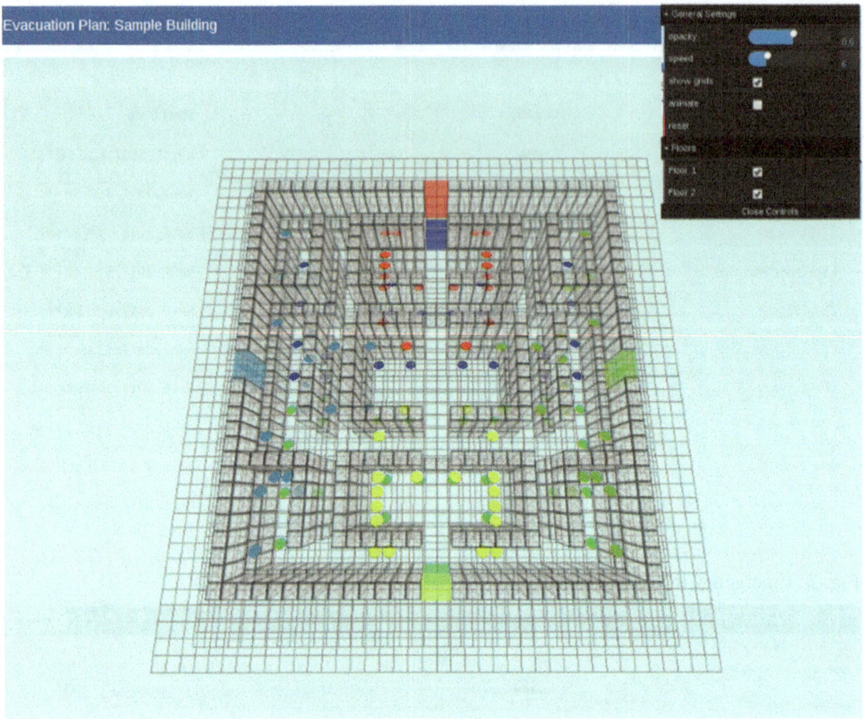

Fig. 4 Evacuation simulation component

4 Conclusions

In this paper, we presented a multiagent based system that provides real-time emergency management and analysis by utilizing microblog data streams. It consists of commonly used classification and clustering approaches for filtering useful data. It has the ability to detect location and extract additional information that summarizes the event from social media posts. In case of an emergency in a place that has architectural plan stored in the system, optimal evacuation plan is simulated in the user interface. Development of the system revealed that, if it is used the right way, social media analysis can help us to detect and monitor disasters and other emergencies in real time as early as possible. As a future work, we plan to perform more information extraction for a better summary of the events. We will try to reflect lessons learned from one event to better handle emerging events. This could be possible by capturing, integrating, storing, and analyzing emergency data from various sources. In addition, we will create a framework that can be used in partially observable rescue scenarios. It will help us to come up with more realistic solutions when the entire environment data is not available in every time-slot.

References

1. A. Sadilek, H. Kautz, V. Silenzio, Predicting disease transmission from geo-tagged micro-blog data, in *Proceedings of the Twenty-Sixth AAAI Conference on Artificial Intelligence*, Series AAAI'12 (AAAI Press, Toronto, 2012), pp. 136–142
2. A. Lamb, M.J. Paul, M. Dredze, Separating fact from fear: tracking flu infections on twitter, in *NAACL* (2013)
3. Y. Kryvasheyeu, H. Chen, N. Obradovich, E. Moro, P. Van Hentenryck, J. Fowler, M. Cebrian, Rapid assessment of disaster damage using social media activity. Sci. Adv. **2**(3), e1500779, 03 (2016)
4. A. Gupta, P. Kumaraguru, *Misinformation in Social Networks, Analyzing Twitter During Crisis Events* (Springer New York, New York, NY, 2014), pp. 922–931
5. M. Avvenuti, M.G. Cimino, S. Cresci, A. Marchetti, M. Tesconi, A framework for detecting unfolding emergencies using humans as sensors. Springerplus **5**, 43 (2016)
6. R. McCreadie, C. Macdonald, I. Ounis, Eaims: emergency analysis identification and management system, in *Proceedings of the 39th International ACM SIGIR Conference on Research and Development in Information Retrieval*, Series SIGIR '16 (ACM, New York, NY, 2016), pp. 1101–1104
7. R. McCreadie, C. Macdonald, I. Ounis, M. Osborne, S. Petrovic, Scalable distributed event detection for twitter, in *2013 IEEE International Conference on Big Data* (IEEE, New York, NY, 2013), pp. 543–549
8. M. Dredze, M.J. Paul, S. Bergsma, H. Tran, Carmen: a twitter geolocation system with applications to public health (2013)
9. C. Castillo, M. Mendoza, B. Poblete, Information credibility on twitter, in *Proceedings of the 20th International Conference on World Wide Web* (ACM, New York, 2011), pp. 675–684
10. D. Kanojia, V. Kumar, K. Ramamritham, Civique: using social media to detect urban emergencies. CoRR **abs/1610.04377** (2016)
11. M. Avvenuti, F.D. Vigna, S. Cresci, A. Marchetti, M. Tesconi, Pulling information from social media in the aftermath of unpredictable disasters, in *2015 2nd International Conference on Information and Communication Technologies for Disaster Management (ICT-DM)*, Nov 2015, pp. 258–264
12. S. Verma, S. Vieweg, W.J. Corvey, L. Palen, J.H. Martin, M. Palmer, A. Schram, K.M. Anderson, Natural language processing to the rescue? extracting "situational awareness" tweets during mass emergency, in *ICWSM* (2011)
13. T. Sakaki, M. Okazaki, Y. Matsuo, Earthquake shakes twitter users: real-time event detection by social sensors, in *Proceedings of the 19th International Conference on World Wide Web* (ACM, New York, 2010), pp. 851–860
14. A.-M. Popescu, M. Pennacchiotti, D. Paranjpe, Extracting events and event descriptions from twitter, in *Proceedings of the 20th International Conference Companion on World Wide Web*, Series WWW '11 (ACM, New York, 2011), pp. 105–106
15. G.I. Hawe, G. Coates, D.T. Wilson, R.S. Crouch, Agent-based simulation of emergency response to plan the allocation of resources for a hypothetical two-site major incident. Eng. Appl. Artif. Intell. **46**(Part B), 336–345 (2015)
16. M. Bandyopadhyay, V. Singh, Development of agent based model for predicting emergency response time. Perspect. Sci. **8**, 138–141 (2016)
17. R. Safdari, J.S. Malak, N. Mohammadzadeh, A.D. Shahraki, A multi agent based approach for prehospital emergency management. Bull. Emerg. Trauma **5**, 171–178 (2017)
18. B. Lopez, B. Innocenti, D. Busquets, A multiagent system for coordinating ambulances for emergency medical services. IEEE Intell. Syst. **23**, 50–57 (2008)
19. M.N.R. Safdari, A. Rahimi, Multi-agent system as a new approach to effective chronic heart failure management: key considerations. Healthc. Informatics Res. **19**, 162–166 (2013)
20. T. Simon, A. Goldberg, B. Adini, Socializing in emergencies-a review of the use of social media in emergency situations. Int. J. Inf. Manag. **35**(5), 609–619 (2015)

21. A. Gupta, H. Lamba, P. Kumaraguru, A. Joshi, Faking sandy: characterizing and identifying fake images on twitter during hurricane sandy, in *Proceedings of the 22Nd International Conference on World Wide Web*, Series WWW '13 Companion (ACM, New York, NY, 2013), pp. 729–736
22. R. Gonzalez, G. Figueroa, Y.-S. Chen, Tweolocator: a non-intrusive geographical locator system for twitter, in *Proceedings of the 5th ACM SIGSPATIAL International Workshop on Location-Based Social Networks*, Series LBSN '12 (ACM, New York, NY, 2012), pp. 24–31
23. M. Oussalah, F. Bhat, K. Challis, T. Schnier, A software architecture for twitter collection, search and geolocation services. Knowl. Based Syst. **37**, 105–120 (2013)
24. H.S. Burton, W.K. Tanner, G.C. Giraud-Carrier, H.J. West, D.M. Barnes, right time, right place health communication on twitter: value and accuracy of location information. J. Med. Internet Res. **14**(6), e156 (2012)
25. J. Piskorski, R. Yangarber, *Information Extraction: Past, Present and Future* (Springer, Berlin, 2013), pp. 23–49
26. J.R. Finkel, T. Grenager, C. Manning, Incorporating non-local information into information extraction systems by gibbs sampling, in *Proceedings of the 43rd Annual Meeting on Association for Computational Linguistics*, Series ACL '05 (Association for Computational Linguistics, Stroudsburg, PA, 2005), pp. 363–370
27. S. Bird, E. Klein, E. Loper, *Natural Language Processing with Python* (O'Reilly Media, Beijing, 2009)
28. X. Pan, C.S. Han, K. Dauber, K.H. Law, A multi-agent based framework for the simulation of human and social behaviors during emergency evacuations. AI Soc. **22**(2), 113–132 (2007)

Social Network Analysis to Combat Terrorism: 2015 Paris Attacks

Animesh Gupta, Tansel Özyer, Jon Rokne, and Reda Alhajj

Abstract A series of attacks shook Paris in the year 2015. They were well-coordinated attacks by the terrorist organization ISIL. An in-depth analysis of the attacks is presented in this chapter. This work is divided into several stages and every stage progresses into the next stage by adding more information useful for the upcoming stage. Initially, data about terrorists is gathered from newspapers and online bulletins. The collected data is then transformed into a network which is created using an adjacency matrix. Strength of the relationship between the involved terrorists is also factored in while creating the network. To gather insight into this network, centrality measures are calculated. This analysis brought forward interesting facts such as who were the most important person(s) in the network and removing which person(s) would cripple the network. The final phase of this process was a Twitter analysis of the Paris attacks based on four keywords related to the attack. This further revealed a few facts which were unreported in major newspapers. Some conclusions and understanding of the attacks after an in-depth data analysis of Paris attacks are also reported.

1 Introduction

The 2015 Paris attacks were a series of coordinated terrorist attacks that occurred on Friday, November 13, 2015, in Paris, France, and the city's northern suburb, Saint-Denis [1]. The attacks started with three suicide bombers wearing suicide vests striking outside the Stade de France stadium in Saint-Denis while there was

---

A. Gupta (✉) · J. Rokne · R. Alhajj
Department of Computer Science, University of Calgary, Calgary, AB, Canada
e-mail: gupta@ucalgary.ca; rokne@ucalgary.ca; alhajj@ucalgary.ca

T. Özyer
Department of Computer Engineering, TOBB University of Economics and Technology, Ankara, Turkey
e-mail: ozyer@etu.edu.tr

T. Özyer et al. (eds.), *Social Networks and Surveillance for Society*, Lecture Notes in Social Networks, https://doi.org/10.1007/978-3-319-78256-0_10

an ongoing soccer match between France and Germany. The attack at the stadium was followed by various bombings and shootings across the city at various cafés and restaurants. Also, three terrorists entered the Bataclan theatre during a concert and started shooting randomly at people in the theatre. The attack at Bataclan caused the highest number of causalities. Hostages were then taken at the Bataclan theatre followed by a standoff between the police and the terrorists. The standoff ended with all the attackers being shot down by the police.

130 people were killed in the Paris attacks, 89 of whom were killed at the Bataclan. Out of 352 people injured, 99 were critically injured [2]. The horrific incident was a series of seven attacks by nine terrorists. Eight terrorists were killed and one was captured by the police.

On November 14, ISIL—the so-called Islamic State of Iraq and the Levant—claimed responsibility for the attacks. The attacks, as claimed in the video, were an act of retaliation by ISIL on the French Government for their airstrikes on Syria and Iraq [3].

Figure 1 shows the locations of the attacks on a map of Paris. French police raided an apartment early on November 18 in the northern Paris suburb of Saint-Denis, a few miles from where the suicide bomb attacks at the Stade de France took place. Three people died, including Abdelhamid Abaaoud, a 28-year-old Belgian man who is suspected of planning the Paris attacks, and his 26-year-old cousin Hasna Aitboulahcen [4].

The aim of the study described in this chapter is to analyze Paris attacks in a scientific way and draw lessons which may help to avoid future similar disasters. For this purpose, we employ social network construction and analysis in the process. Indeed, as homeland security is concerned, social network analysis may be considered as a mathematical modelling technique for connecting the dots and using science and statistics to fight criminology and terrorism [5]. The research on social network analysis (SNA) to combat terrorism has propelled since the occurrence of September 11, 2001, attacks in New York, USA. Since then, there has been an increased research interest to gather insights into terror organizations but there is still room for further research to yield great insights (such as importance, power, influence, behavior) which can predict and possibly prevent a terror attack. Thus, the objective of this research paper is to gather information about the terrorist network of Paris attacks from some of the publicly available datasets and create useful insights based on node measures, network closures, and Twitter analysis. The target is to draw some conclusions which may guide investigators to may be pre-act and avoid disasters by better handling and coping with similar cases.

The rest of this chapter is organized as follows. Section 2 briefly covers related work. Section 3 describes the methodology applied in this study; it also includes results which show the importance of various terrorists who participated in Paris attacks. Section 4 reports analysis results of tweets related to Paris attacks; and finally some conclusions are presented.

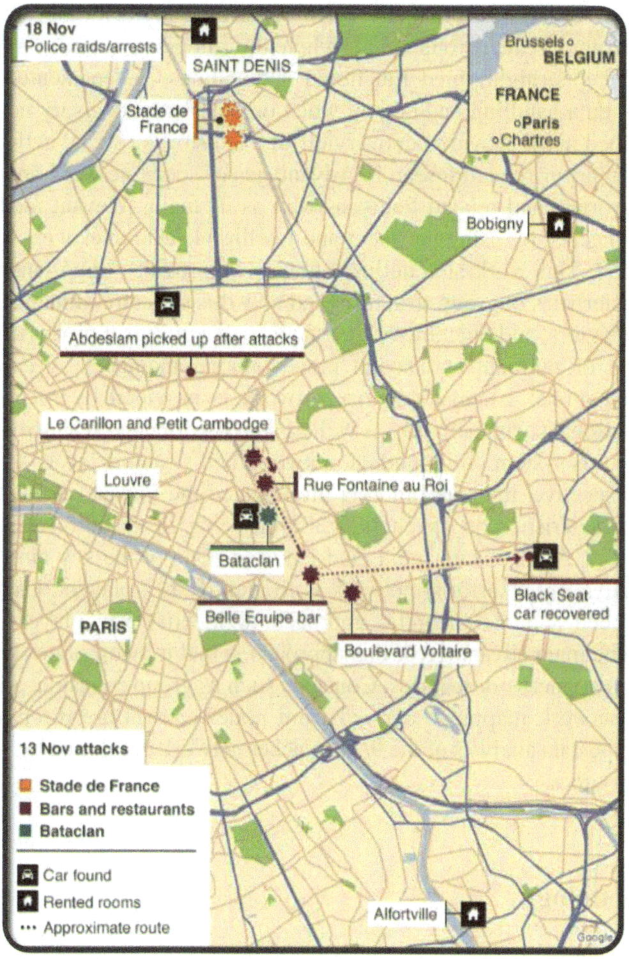

Fig. 1 Map of Paris showing attack locations and other key information

2 Related Work

There has been a constant battle against terrorism and many articles have been published by various research groups and contributed towards the growing research to combat terrorism. For instance, Koschade [6] attempted to advance the growing literature on social network analysis and terrorism studies through a social network analysis of the Jemaah Islamiyah cell that was responsible for the Bali bombings in 2002. The aim of his research was to understand the communication within and structure of such cells and assist to predict the outcomes of terrorist cells. A general introduction to social network analysis (SNA) as an analytical tool for the study of adversary networks can be found in [7] where the authors reviewed some theoretical and key concepts, highlighted functional applications, and presented a tentative protocol for data handling and coding when dealing with adversary networks. The work described in [8] provides an overview of the history of social network analysis and its use in terror-related research. Advantages of social network analysis for the study of terrorism and related fields, as well as its main, relevant methodological tools and concepts, by using pertinent and intelligible examples were discussed by Arie et al. [9]. This work also outlined how SNA provides important information about characteristics of group structure (and how this structure influences members' motives, behaviors, and outcome of their actions), recruitment processes, evolution, and division of political and social power among members. A case study on the so-called global Islamist terrorist network is provided by Medina [10]; he utilized traditional social network, as well as small-world, and scale-free analyses to characterize this system on individual, network, and systemic levels. From a social network perspective, the study in [11] investigated the impact of expatriate social network characteristics on psychological well-being in the terrorism-endangered environment of Afghanistan, India, and Pakistan. One of the key aims of a social network analysis of a terrorist network is to identify key players. Lindelauf et al. [12] introduced a game theoretic approach to identify key players in terrorist networks. The main advantage of this approach is that it incorporates both network structure and nonnetwork features. Concepts such as group cohesion, adhesion, and alternative network mappings derived from node removal are discussed in [13], inspired by the data analysis of the 9/11 hijacker network developed by Valdis Krebs from open sources.

3 Methodology

The following figure shows the five stages into which the first phase of the work described in this chapter has been divided. The second phase is the Twitter analysis of the Paris attacks. The five stages of the first phase are described further in this section.

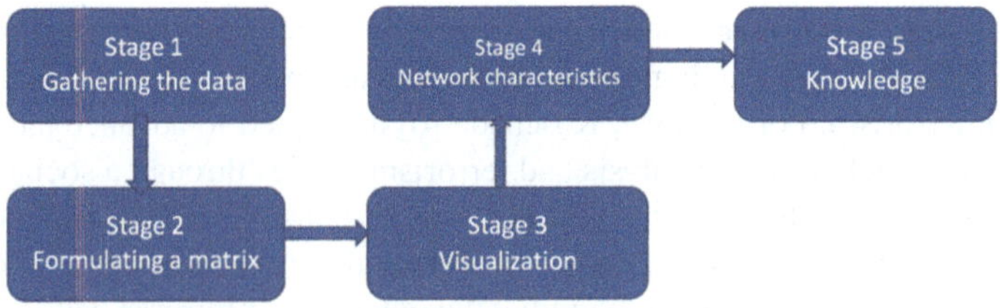

Stage 1: Gathering the Data

In the data-gathering stage, information was gathered from articles and reports on Paris attacks published by major newspapers and websites starting from November 2015 till the date this chapter was completed. For this stage, we have been able to come up with a list of 21 terrorists who were either actively or passively involved in the attacks. By active participation, we mean terrorists who wore suicide vests and blew themselves up or terrorists who shot down people at Bataclan. Passive participation involves people who were not present at the scene; these may include drivers, recruiters, bomb makers, etc. Here is the list of all the 21 terrorists:

1. Bilal Hadfi
2. Ahmad al-Mohammed
3. M al Mahmod
4. Chakib Akrouh
5. Abdelhamid Abaaoud
6. Brahim Abdeslam
7. Omar Ismail Mostefai
8. Samy Aminour

Active Participation

9. Foued Mohamed-Aggad
10. Salah Abdeslam
11. Mohamed Abrini
12. Hasna Ait Boulahcen
13. Fabien Clain
14. Mehdi Nemmoude
15. Mourad Fares
16. Jawad Bendaoud
17. Mohammed Amri
18. Hamza Attouh
19. Abraimi Lazer
20. Abou Isleym

Passive Participation

21. Gelal Attar

The first 9 terrorists in the list are those who had an active participation in the attacks whereas the remaining 12 terrorists are those who had a passive participation. This list will now act as a dataset to form an adjacency matrix to define the relationships between these terrorists. The outcome will feed the following stages in the process.

Stage 2: Formulating a Matrix

A suspect/terrorist may have many contacts—both accidental and intentional [14]—but not all terrorists have the same strength of relationships with other terrorists. As data comes in, a picture of the terrorist organization slowly comes into focus [14]. Some share very strong bonds being family members, whereas others share a weak bond if they communicated just once over a phone call. A weighted adjacency matrix is created based on the strength of the relationships as follows:

Weight 1 corresponds to a very weak relationship, weight 2 corresponds to an intermediate relationship, and weight 3 corresponds to a very strong relationship between terrorists.

The adjacency matrix is a 21×21 matrix with terrorist names forming rows and columns of the matrix, and the strength of the relationship on a scale of 1–3 is used to decide on values of the entries in the matrix. The matrix has been appended at the end of the chapter.

Stage 3: Visualization

In this stage, we visualize the data by creating a network from the adjacency matrix. This has been done using R. Nodes in the network correspond to terrorists who were involved in the attacks and edges exist between terrorists sharing a relationship—weak or strong. Also, each edge is marked with a strength based on the weight matrix shown in Table 1.

Terrorist networks are very secretive and focus primarily on being clandestine than being efficient. These networks rely on moderators to act as a bridge between people who exist in the network. Moderators are not an actual part of the network, but appear briefly in the network during meetings and briefings. During the short duration of their presence, they act as a means of communication between other members of the group and then vanish until the next meeting. Therefore, the graph we obtained is a sparse network and the actors involved are not well connected with each other even after being in the same team. Most people who were involved did not know each other before the attacks. They were brought into contact only for the attack, and are therefore connected only through Jihadist recruiters in the network.

Table 1 Weight matrix

Weight	Type of interaction
1	Single text/transaction/phone call
2	Drove/recruiter
3	Family/prison mates/lived together

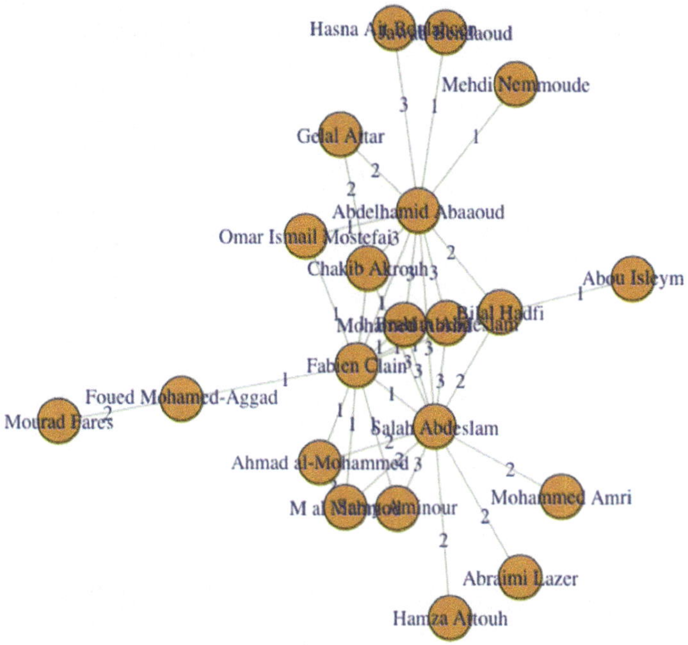

Fig. 2 Network of terrorists involved in the Paris attacks

In the network shown in Fig. 2, Abdelhamid Abaaoud, Salah Abdeslam, and Fabien Clain are the most densely connected terrorists. Newspaper reports claimed that Abdelhamid Abaaoud was the mastermind of this attack and our network supports this claim. Once we have the centrality measures for all the nodes, we should be able to explain and justify why Salah Abdeslam and Fabien Clan are densely connected in the network.

Stage 4: Network Characteristics

In this stage, we used NetDriller [15] and calculated the centrality measures for the graph created in the previous stage. This helped us to answer some implicit questions about the network which are not so evident from just looking at the facts. The four centrality measures which we have calculated are the following:

- *Degree Centrality*

It is defined as the number of links incident upon a node, i.e., the number of ties that a node has.

Table 2 Degree centrality

Degree centrality	
Node name	Score
Salah Abdeslam	12
Fabien Clain	11
Abdelhamid Abaaoud	11
Chakib Akrouh	4
Bilal Hadfi	4
M al Mahmod	3
Ahmad al-Mohammed	3
Brahim Abdeslam	3
Mohamed Abrini	3
Gelal Attar	2
Hasna Ait Boulahcen	2
Omar Ismail Mostefai	2
Samy Aminour	2
Jawad Bendaoud	2
Foued Mohamed-Aggad	2
Abou Isleym	1
Mohammed Amri	1
Abraimi Lazer	1
Mourad Fares	1
Hamza Attouh	1

- *Betweenness Centrality*

Betweenness centrality quantifies the number of times a node acts as a bridge along the shortest path between two other nodes.

- *Closeness Centrality*

Closeness centrality of a node is the average length of the shortest paths between the node and all other nodes in the graph.

- *Eigenvector Centrality*

Eigenvector centrality is a measure of the influence of a node in a network.

Tables 2–5 show these four centrality measures, respectively, for all terrorists. These measures helped us to fill the voids and answer questions which are not evident from a superficial study. We will now discuss these findings in the next stage.

Table 3 Betweenness centrality

Betweenness centrality	
Node name	Score
Salah Abdeslam	73
Abdelhamid Abaaoud	70.5
Fabien Clain	62.5
Bilal Hadfi	19
Foued Mohamed-Aggad	19
Chakib Akrouh	5
Gelal Attar	0
Abou Isleym	0
Hasna Ait Boulahcen	0
Mohammed Amri	0
Abraimi Lazer	0
Omar Ismail Mostefai	0
M al Mahmod	0
Mourad Fares	0
Ahmad al-Mohammed	0
Samy Aminour	0
Hamza Attouh	0
Brahim Abdeslam	0
Mehdi Nemmoude	0
Jawad Bendaoud	0

Table 4 Closeness centrality

Closeness centrality	
Node name	Score
Salah Abdeslam	0.689655
Fabien Clain	0.689655
Abdelhamid Abaaoud	0.666666
Bilal Hadfi	0.54054
Chakib Akrouh	0.526315
Brahim Abdeslam	0.51282
Mohamed Abrini	0.51282
Omar Ismail Mostefai	0.465116
M al Mahmod	0.465116
Ahmad al-Mohammed	0.465116
Samy Aminour	0.454545
Foued Mohamed-Aggad	0.434782
Gelal Attar	0.416666
Hasna Ait Boulahcen	0.416666
Mohammed Amri	0.416666
Abraimi Lazer	0.416666
Hamza Attouh	0.416666
Jawad Bendaoud	0.416666
Mehdi Nemmoude	0.408163
Abou Isleym	0.357142

Table 5 Eigenvector centrality

Eigenvector centrality	
Node name	Score
Salah Abdeslam	0.166666
Abdelhamid Abaaoud	0.152777
Fabien Clain	0.152777
Bilal Hadfi	0.055555
Chakib Akrouh	0.055555
Brahim Abdeslam	0.041666
Mohamed Abrini	0.041666
M al Mahmod	0.041666
Ahmad al-Mohammed	0.041666
Foued Mohamed-Aggad	0.027777
Samy Aminour	0.027777
Omar Ismail Mostefai	0.027777
Hasna Ait Boulahcen	0.027777
Jawad Bendaoud	0.027777
Gelal Attar	0.027777
Mohammed Amri	0.013888
Abraimi Lazer	0.013888
Hamza Attouh	0.013888
Mehdi Nemmoude	0.013888
Abou Isleym	0.013888

Stage 5: Knowledge Discovery

Network characteristics in terms of centrality measures reveal a lot of information about the terrorists who were involved in Paris attacks. Salah Abdeslam, Fabien Clan, and Abdelhamid Abaaoud have the highest centrality measures across all four parameters. These results are consistent with the fact that Abdelhamid Abaaoud, who was killed at Saint-Denis by the French police 5 days after the attack, was the mastermind of Paris attacks. Although he does not have the highest value in our calculations he is placed among the top three in each of the centrality measures. Abdelhamid Abaaoud not being on top in the centrality measures can be attributed to some missing or incorrect data which is common in a clandestine network analysis. Salah Abdeslam is the terrorist who has the highest centrality measures across all four parameters. He was arrested by French police during the attacks and is still in custody. According to the centrality measures, he was well connected within the network and can provide intelligence related to the attacks and the attackers which will be useful to understand the dynamics of this network in terms of evolution of the network by the addition of new members or the formation of new relationships between existing members.

Bataclan attackers, even if the death toll was the highest, seem to have had a much more peripheral role in the operation. Salah Abdeslam also had an important role in coordinating the different attacks [17]. Also, several links between Abdeslam and the suspected Paris attacks' mastermind, Abdelhamid Abaaoud, emerged in the days after the attacks [18]. Fabien Clan is believed to be one of the recruiters for Paris attacks and is the third most important person according to the centrality measures. He is also the voice in the video which was released by ISIL after the attacks claiming responsibility for the attacks. Although he was not directly involved in the attacks he had strong connections with most of the people who were involved in the attacks. Early detection of a person like Fabien Clan could have disrupted the entire network and possibly could have prevented an attack of this destructive magnitude.

4 Twitter Analysis

The following analysis was performed on a machine running Windows 7, 64-bit operating system, and having 8 GB of RAM.

In [16], Moujahid gave an overview of how text mining can be performed using Twitter streaming API and Python. He used Twitter data as an example to compare the popularity of three programming languages, Python, Javascript, and Ruby, and to retrieve links to programming tutorials. Some parts of this work are inspired from Adil's article.

For this phase, python is used for the analysis of tweets which are extracted from a dataset containing twitter IDs of tweets corresponding to the four keywords—#parisattacks, #Bataclan, #paris, and #porteouverte. Hydrating the tweets gave us 6.5 million tweets. These tweets formed the basis for our analysis. To be able to predict an attack, the time when the tweets about the attack start rolling in is of utmost importance. Therefore, we extracted the timestamp for all the tweets as our most important feature to work with.

Since we know that the first attack (from the series of attacks) occurred at 9:16 PM on July 14, we construct a timeline starting from 8 PM on July 14 until 1 AM on July 15. Each one of the 6.5 million tweets is then mapped to this timeline based on the time when the tweet was received. Figure 3 shows the timeline with the corresponding number of tweets for every second within the chosen time period.

Although the attacks started at 9:16 PM, there is an absence of significant number of tweets between 8 PM and 9:30 PM. The magnitude of the tweets started increasing from 9:30 PM onwards and reached a maximum around 10:15 PM. There is a steady number of tweets after that till 12:30 AM after which the number of tweets starts decreasing again. This is shown graphically in Fig. 3 in which the height of the histogram at any time is proportional to the number of tweets at that time.

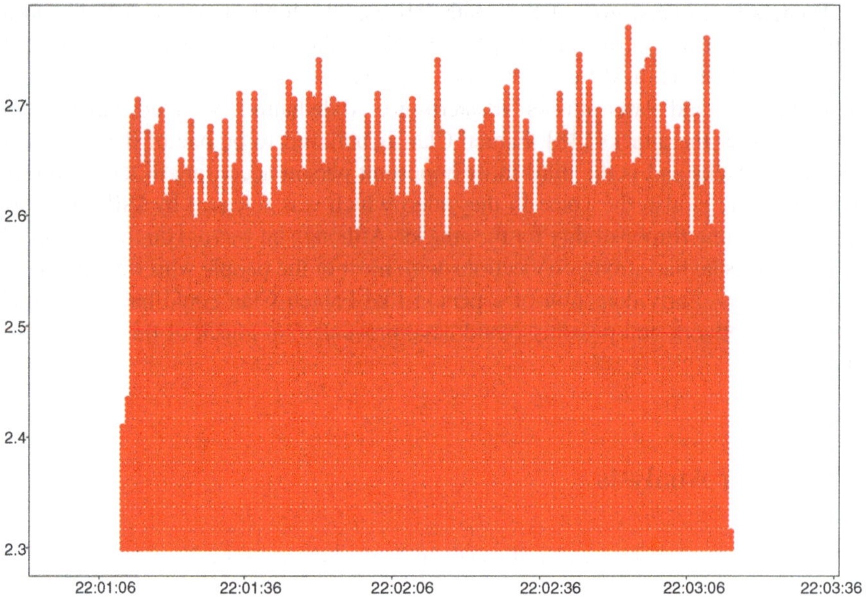

Fig. 3 Tweet timeline for Paris attacks

This work can be an essential cornerstone for an early detection system for a terrorist attack. Although predicting an attack of this magnitude is impossible based solely on twitter data because of the highly secretive nature of terrorist organizations, an early detection system is a need of the hour. An early detection system can detect an attack just a few minutes after any occurrence so that any subsequent occurrences can be avoided or tacked with the gained intelligence. The results and observations highlighted in this chapter about the terrorist network of Paris attacks can be used to base such a model on. The organization and working of any terrorist network will not be much different from the terrorist network explained in this chapter as all these organizations have the same goal and work according to a similar principle.

Appendix 1: Adjacency matrix

	Bilal Hadfi	Ahmad al-Mohammed	M al Mahmod	Chakib Akrouh	Abdelhamid Abaaoud	Brahim Abdeslam	Omar Ismail Mostefai	Samy Aminour	Foued Mohamed-Aggad	Salah Abdeslam
Bilal Hadfi	0	0	0	0	2	0	0	0	0	2
Ahmad al-Mohammed	0	0	2	0	0	0	0	0	0	2
M al Mahmod	0	2	0	0	0	0	0	0	0	2
Chakib Akrouh	0	0	0	0	3	0	0	0	0	3
Abdelhamid Abaaoud	2	0	0	3	0	3	1	0	0	3
Brahim Abdeslam	0	0	0	0	3	0	0	0	0	0
Omar Ismail Mostefai	0	0	0	0	1	0	0	0	0	0
Samy Aminour	0	0	0	0	0	0	0	0	0	3
Foued Mohamed-Aggad	0	0	0	0	0	0	0	0	0	0
Salah Abdeslam	2	2	2	3	3	3	0	3	0	0
Mohamed Abrini	0	0	0	0	3	0	0	0	0	3
Hasna Ait Boulahcen	0	0	0	0	3	0	0	0	0	0
Fabien Clain	1	1	1	1	1	1	1	1	1	1
Mehdi Nemmoude	0	0	0	0	1	0	0	0	0	0
Mourad Fares	0	0	0	0	0	0	0	0	2	0
Jawad Bendaoud	0	0	0	0	1	0	0	0	0	0
Mohammed Amri	0	0	0	0	0	0	0	0	0	2
Hamza Attouh	0	0	0	0	0	0	0	0	0	2
Abraimi Lazer	0	0	0	0	0	0	0	0	0	2
Abou Isleym	1	0	0	0	0	0	0	0	0	0
Gelal Attar	0	0	0	2	2	0	0	0	0	0

	Mohamed Abrini	Hasna Ait Boulahcen	Fabien Clair	Mehdi Nemmoude	Mourad Fares	Jawad Bendaoud	Mohammed Amri	Hamza Attouh	Abraimi Later	Abou Isleym	Gelal Attar
Bilal Hadfi	0	0	1	0	0	0	0	0	0	1	0
Ahmad al-Mohammed	0	0	1	0	0	0	0	0	0	0	0
M al Mahmod	0	0	1	0	0	0	0	0	0	0	0
Chakib Akrouh	0	0	1	0	0	0	0	0	0	0	2
Abdelhamid Abaaoud	3	3	1	1	0	1	0	0	0	0	2
Brahim Abdeslam	0	0	1	0	0	0	0	0	0	0	0
Omar Ismail Mostefai	0	0	1	0	0	0	0	0	0	0	0
Samy Aminour	0	0	1	0	0	0	0	0	0	0	0
Foued Mohamed-Aggad	0	0	1	0	2	0	0	0	0	0	0
Salah Abdeslam	3	0	1	0	0	0	2	2	2	0	0
Mohamed Abrini	0	0	1	0	0	0	0	0	0	0	0
Hasna Ait Boulahcen	0	0	0	0	0	1	0	0	0	0	0
Fabien Clain	1	0	0	0	0	0	0	0	0	0	0
Mehdi Nemmoude	0	0	0	0	0	0	0	0	0	0	0
Mourad Fares	0	0	0	0	0	0	0	0	0	0	0
Jawad Bendaoud	0	1	0	0	0	0	0	0	0	0	0
Mohammed Amri	0	0	0	0	0	0	0	0	0	0	0
Hamza Attouh	0	0	0	0	0	0	0	0	0	0	0
Abraimi Lazer	0	0	0	0	0	0	0	0	0	0	0
Abou Isleym	0	0	0	0	0	0	0	0	0	0	0
Gelal Attar	0	0	0	0	0	0	0	0	0	0	0

References

1. J. Kostov, France honors victims of Paris terrorist attacks. WSJ. (2017). https://www.wsj.com/articles/france-honors-victims-of-paris-terrorist-attacks-1448635450
2. Injuries from Paris attacks will take long to heal. Cbsnews.com. (2017). https://www.cbsnews.com/news/injuries-from-paris-attacks-will-take-long-to-heal/
3. France, Russia strike Islamic state; new suspect sought. Reuters. (2017). https://www.reuters.com/article/us-france-shooting-idUSKCN0T22IU20151117
4. Unravelling the connections among the Paris attackers. Nytimes.com. (2017). https://www.nytimes.com/interactive/2015/11/15/world/europe/manhunt-for-paris-attackers.html?_r=0
5. A. Afonso, Social network analysis of terrorist networks. Orgnet.com. (2017). http://www.orgnet.com/hijackers.html
6. S.A. Koschade, A social network analysis of Jemaah Islamiyah: the applications to counter-terrorism and intelligence. Stud. Conflict Terror **29**(6), 559–575 (2006)
7. R.C. van der Hulst, Introduction to social network analysis (SNA) as an investigative tool. Trends Organ Crime **12**(2), 101–121 (2009)
8. Ressler, S., Social network analysis as an approach to combat terrorism: past, present, and future research, homeland security affairs 2, (2006)
9. A. Perliger, A. Pedahzur, Social network analysis in the study of terrorism and political violence. Political Sci Politics **44**(1), 45–50 (2011)
10. R. Medina, Social network analysis: a case study of the Islamist terrorist network. Secur. J. **27**(1), 97–121 (2012)
11. B. Bader, T. Schuster, Expatriate social networks in terrorism-endangered countries: an empirical analysis in Afghanistan, India, Pakistan, and Saudi Arabia. J. Int. Manag. **21**(1), 63–77 (2015)
12. R. Lindelauf, H. Hamers, B. Husslage, Game theoretic centrality analysis of terrorist networks: the cases of Jemaah Islamiyah and Al Qaeda, CentER Discussion Paper Series No. 2011–107, (2011)
13. P. Fellman, R. Wright, Modeling terrorist networks, complex systems at the mid-range, the intelligencer. J. U.S. Int. Stud. **14**(1), (2014)
14. Connecting the Dots—Social network analysis of 9–11 terror network. Orgnet.com. (2017). http://www.orgnet.com/tnet.html
15. N. Koochakzadeh, A. Sarraf, K. Kianmehr, J.G. Rokne, R. Alhajj, Netdriller: a powerful social network analysis tool, in *Proceedings of IEEE ICDM*. 1235–1238. (2011)
16. An introduction to text mining using twitter streaming API and Python//Adil Moujahid//Data analytics and more. Adilmoujahid.com. (2017). http://adilmoujahid.com/posts/2014/07/twitter-analytics/
17. Web of terror from Paris to Brussels. Mail Online. (2017). http://www.dailymail.co.uk/news/article-3506304/Web-terror-Paris-Brussels.html
18. J. Klausen, Tweeting the Jihad: social media networks of western foreign fighters in Syria and Iraq. Stud Conflict and Terror **38**(1), 1–22 (2014)

The manufacturer's authorised representative in the EU is Springer Nature Customer Service Centre GmbH, Europaplatz 3, 69115 Heidelberg, Germany. If you have any concerns regarding our products, please contact ProductSafety@springernature.com

Printed and bound by CPI Group (UK) Ltd, Croydon, CR0 4YY
30/09/2026
02225658-0019